DARWIN'S BLACK BOX

THE BIOCHEMICAL CHALLENGE TO EVOLUTION

MICHAEL J. BEHE

A TOUCHSTONE BOOK PUBLISHED BY
SIMON & SCHUSTER

TOUCHSTONE
Rockefeller Center
1230 Avenue of the Americas
New York, NY 10020

First Touchstone Edition 1998

TOUCHSTONE and colophon are registered trademarks
of Simon & Schuster Inc.

Designed by Carla Bolte

Manufactured in the United States of America

10   9   8   7   6

Library of Congress Cataloging-in-Publication Data

Behe, Michael J.
   Darwin's black box : the biochemical challenge to evolution /
Michael J. Behe.
     p. cm.
   Includes bibliographical references and index.
   ISBN 0-684-82754-9
       0-684-83493-6 (Pbk)
   1. Molecular evolution. 2. Evolution (Biology) I. Title.
QH325.B365 1996
575—dc20   96-695
                                  CIP

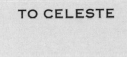

TO CELESTE

CONTENTS

PREFACE

A MOLECULAR PHENOMENON

It is commonplace, almost banal, to say that science has made great strides in understanding nature. The laws of physics are now so well understood that space probes fly unerringly to photograph worlds billions of miles from earth. Computers, telephones, electric lights, and untold other examples testify to the mastery of science and technology over the forces of nature. Vaccines and high-yield crops have stayed the ancient enemies of mankind, disease and hunger—at least in parts of the world. Almost weekly, announcements of discoveries in molecular biology encourage the hope of cures for genetic diseases and more.

Yet understanding how something works is not the same as understanding how it came to be. For example, the motions of the planets in the solar system can be predicted with tremendous accuracy; however, the origin of the solar system (the question of how the sun, planets, and their moons formed in the first place) is still controversial.[1] Science may eventually solve the riddle. Still, the point remains that understanding the origin of something is different from understanding its day-to-day workings.

Science's mastery of nature has led many people to presume that it can—indeed, must—also explain the origin of nature and life. Darwin's proposal that life can be explained by natural selection acting on variation has been overwhelmingly accepted in educated circles for more than a century, even though the basic mechanisms of life remained utterly mysterious until several decades ago.

Modern science has learned that, ultimately, life is a molecular phenomenon: All organisms are made of molecules that act as the nuts and bolts, gears and pulleys of biological systems. Certainly there are complex biological features (such as the circulation of blood) that emerge at higher levels, but the gritty details of life are the province of biomolecules. Therefore the science of biochemistry, which studies those molecules, has as its mission the exploration of the very foundation of life.

Since the mid-1950s biochemistry has painstakingly elucidated the workings of life at the molecular level. Darwin was ignorant of the reason for variation within a species (one of the requirements of his theory), but biochemistry has identified the molecular basis for it. Nineteenth-century science could not even guess at the mechanism of vision, immunity, or movement, but modern biochemistry has identified the molecules that allow those and other functions.

It was once expected that the basis of life would be exceedingly simple. That expectation has been smashed. Vision, motion, and other biological functions have proven to be no less sophisticated than television cameras and automobiles. Science has made enormous progress in understanding how the chemistry of life works, but the elegance and complexity of biological systems at the molecular level have paralyzed science's attempt to explain their origins. There has been virtually no attempt to account for the origin of specific, complex biomolecular systems, much less any progress. Many scientists have gamely asserted that explanations are already in hand, or will be sooner or later, but no support for such assertions can be found in the professional science literature. More importantly, there are compelling reasons—based on the structure of the systems themselves—to think that a Darwinian explanation for the mechanisms of life will forever prove elusive.

Evolution is a flexible word.[2] It can be used by one person to mean something as simple as change over time, or by another person to mean the descent of all life forms from a common ancestor, leaving the mechanism of change unspecified. In its full-throated, biological sense, however, *evolution* means a process whereby life arose from non-living matter and subsequently developed entirely by natural means. That is the sense that Darwin gave to the word, and the meaning that it holds in the scientific community. And that is the sense in which I use the word *evolution* throughout this book.

APOLOGIA FOR DETAILS

Several years ago, Santa Claus gave my oldest son a plastic tricycle for Christmas. Unfortunately, busy man that he is, Santa had no time to take it out of the box and assemble it before heading off. The task fell to Dad. I took the parts out of the box, unfolded the assembly instructions, and sighed. There were six pages of detailed instructions: line up the eight different types of screws, insert two $1\frac{1}{2}$-inch screws through the handle into the shaft, stick the shaft through the square hole in the body of the bike, and so on. I didn't want to even read the instructions, because I knew they couldn't be skimmed like a newspaper—the whole purpose is in the details. But I rolled up my sleeves, opened a can of beer, and set to work. After several hours the tricycle was assembled. In the process I had indeed read every single instruction in the booklet several times (to drill them into my head) and performed the exact actions that the instructions required.

My aversion to instructions seems to be widespread. Although most households own a videocassette recorder (VCR), most folks cannot program them. These technological wonders come with complete operating instructions, but the very thought of tediously studying each sentence of the booklet makes most people delegate the job to the nearest ten-year-old.

Unfortunately, much of biochemistry is like an instruction booklet, in the sense that the importance is in the details. A student of biochemistry who merely skims through a biochemistry textbook is virtually certain to spend much of the next exam staring at the ceiling as drops of sweat trickle down his or her forehead. Skimming the textbook does not prepare a student for questions such as "Outline in detail the mechanism of hydrolysis of a peptide bond by trypsin, paying special attention to the role of transition state binding energy." Although there are broad principles of biochemistry that help a mortal comprehend the general picture of the chemistry of life, broad principles only take you so far. A degree in engineering does not substitute for the tricycle instruction booklet, nor does it directly help you to program your VCR.

Many people, unfortunately, are all too aware of the pickiness of biochemistry. People who suffer with sickle cell anemia, enduring much pain in their shortened lives, know the importance of the little detail that changed one out of 146 amino acid residues in one out of the tens of thousands of proteins in their body. The parents of children who die of Tay-Sachs or cystic fibrosis, or who suffer from diabetes or hemophilia, know more than they want to about the importance of biochemical details.

So, as a writer who wants people to read my work, I face a dilemma: people hate to read details, yet the story of the impact of biochemistry on evolutionary theory rests solely in the details. Therefore I have to write the kind of book people don't like to read in order to persuade them of the ideas that push me to write. Nonetheless, complexity must be experienced to be appreciated. So, gentle reader, I beg your patience; there are going to be a lot of details in this book.

The book is divided into three parts. Part I gives some background and shows why evolution must now be argued at the molecular level—the domain of the science of biochemistry. This portion is largely free from technical details, although some do creep in during a discussion of the eye. Part II contains the "example chapters," where most of the complexity is found. Part III is a nontechnical discussion of the implications of biochemistry's discoveries.

So the hard stuff is confined mostly to Part II. In that section, however, I liberally use analogies to familiar, everyday objects to get the ideas across, and even in that section detailed descriptions of biochemical systems are minimized. Paragraphs that contain the heaviest doses of details—replete with eye-glazing technical terms—are set off from the regular text with the ornament ❐, to brace the reader. Some readers may plow right through Part II. Others, however, may wish to skim the section or even skip parts, then return when they're ready to absorb more. For those who want a deeper understanding of biochemistry, I have included an Appendix outlining some general biochemical principles. I encourage those who want all the details to borrow an introductory biochemistry text from the library.

PART I

THE BOX IS OPENED

CHAPTER 1

LILLIPUTIAN BIOLOGY

THE LIMITS OF AN IDEA

This book is about an idea—Darwinian evolution—that is being pushed to its limits by discoveries in biochemistry. Biochemistry is the study of the very basis of life: the molecules that make up cells and tissues, that catalyze the chemical reactions of digestion, photosynthesis, immunity, and more.[1] The astonishing progress made by biochemistry since the mid-1950s is a monumental tribute to science's power to understand the world. It has brought many practical benefits in medicine and agriculture. We may have to pay a price, though, for our knowledge. When foundations are unearthed, the structures that rest on them are shaken; sometimes they collapse. When sciences such as physics finally uncovered their foundations, old ways of understanding the world had to be tossed out, extensively revised, or restricted to a limited part of nature. Will this happen to the theory of evolution by natural selection?

Like many great ideas, Darwin's is elegantly simple. He observed that there is variation in all species: some members are bigger, some smaller, some faster, some lighter in color, and so forth. He reasoned that since limited food supplies could not support all organisms that

3

are born, the ones whose chance variation gave them an advantage in the struggle for life would tend to survive and reproduce, outcompeting the less favored ones. If the variation were inherited, then the characteristics of the species would change over time; over great periods, great changes might occur.

For more than a century most scientists have thought that virtually all of life, or at least all of its most interesting features, resulted from natural selection working on random variation. Darwin's idea has been used to explain finch beaks and horse hoofs, moth coloration and insect slaves, and the distribution of life around the globe and through the ages. The theory has even been stretched by some scientists to interpret human behavior: why desperate people commit suicide, why teenagers have babies out of wedlock, why some groups do better on intelligence tests than other groups, and why religious missionaries forgo marriage and children. There is nothing—no organ or idea, no sense or thought—that has not been the subject of evolutionary rumination.

Almost a century and a half after Darwin proposed his theory, evolutionary biology has had much success in accounting for patterns of life we see around us. To many, its triumph seems complete. But the real work of life does not happen at the level of the whole animal or organ; the most important parts of living things are too small to be seen. Life is lived in the details, and it is molecules that handle life's details. Darwin's idea might explain horse hoofs, but can it explain life's foundation?

Shortly after 1950 science advanced to the point where it could determine the shapes and properties of a few of the molecules that make up living organisms. Slowly, painstakingly, the structures of more and more biological molecules were elucidated, and the way they work inferred from countless experiments. The cumulative results show with piercing clarity that life is based on *machines*—machines made of molecules! Molecular machines haul cargo from one place in the cell to another along "highways" made of other molecules, while still others act as cables, ropes, and pulleys to hold the cell in shape. Machines turn cellular switches on and off, sometimes killing the cell or causing it to grow. Solar-powered machines capture the energy of photons and store it in chemicals. Electrical machines allow current to flow through nerves. Manufacturing machines build other molecular machines, as well as themselves. Cells swim using machines, copy themselves with machinery, ingest food with machinery. In short, highly sophisticated molecular machines control every cellular process. Thus the details of life are finely calibrated, and the machinery of life enormously complex.

Can all of life be fit into Darwin's theory of evolution? Because the popular media likes to publish exciting stories, and because some scientists enjoy speculating about how far their discoveries might go, it has been difficult for the public to separate fact from conjecture. To find the real evidence you have to dig into the journals and books published by the scientific community itself. The scientific literature reports experiments firsthand, and the reports are generally free of the flights of fancy that make their way into the spinoffs that follow. But as I will note later, if you search the scientific literature on evolution, and if you focus your search on the question of how molecular machines— the basis of life—developed, you find an eerie and complete silence. The complexity of life's foundation has paralyzed science's attempt to account for it; molecular machines raise an as-yet-impenetrable barrier to Darwinism's universal reach. To find out why, in this book I will examine several fascinating molecular machines, then ask whether they can ever be explained by random mutation/natural selection.

Evolution is a controversial topic, so it is necessary to address a few basic questions at the beginning of the book. Many people think that questioning Darwinian evolution must be equivalent to espousing creationism. As commonly understood, creationism involves belief in an earth formed only about ten thousand years ago, an interpretation of the Bible that is still very popular. For the record, I have no reason to doubt that the universe is the billions of years old that physicists say it is. Further, I find the idea of common descent (that all organisms share a common ancestor) fairly convincing, and have no particular reason to doubt it. I greatly respect the work of my colleagues who study the development and behavior of organisms within an evolutionary framework, and I think that evolutionary biologists have contributed enormously to our understanding of the world. Although Darwin's mechanism—natural selection working on variation—might explain many things, however, I do not believe it explains molecular life. I also do not think it surprising that the new science of the very small might change the way we view the less small.

A VERY BRIEF HISTORY OF BIOLOGY

When things are going smoothly in our lives most of us tend to think that the society we live in is "natural," and that our ideas about the world are self-evidently true. It's hard to imagine how other people in other times and places lived as they did or why they believed the things they did. During periods of upheaval, however, when apparently solid verities are questioned, it can seem as if nothing in the world makes sense. During those times history can remind us that the search for reliable knowledge is a long, difficult process that has not yet reached an end. In order to develop a perspective from which we can view the idea of Darwinian evolution, over the next few pages I will very briefly outline the history of biology. In a way, this history has been a chain of black boxes; as one is opened, another is revealed.

Black box is a whimsical term for a device that does something, but whose inner workings are mysterious—sometimes because the workings can't be seen, and sometimes because they just aren't comprehensible. Computers are a good example of a black box. Most of us use these marvelous machines without the vaguest idea of how they work, processing words or plotting graphs or playing games in contented ignorance of what is going on underneath the outer case. Even if we were to remove the cover, though, few of us could make heads or tails of the jumble of pieces inside. There is no simple, observable connection between the parts of the computer and the things that it does.

Imagine that a computer with a long-lasting battery was transported back in time a thousand years to King Arthur's court. How would people of that era react to a computer in action? Most would be in awe, but with luck someone might want to understand the thing. Someone might notice that letters appeared on the screen as he or she touched the keys. Some combinations of letters—corresponding to computer commands—might make the screen change; after a while, many commands would be figured out. Our medieval Englishmen might believe they had unlocked the secrets of the computer. But eventually somebody would remove the cover and gaze on the computer's inner workings. Suddenly the theory of "how a computer works" would be revealed as profoundly naive. The black box that had been slowly decoded would have exposed another black box.

In ancient times *all* of biology was a black box, because no one understood on even the broadest level how living things worked. The ancients who gaped at a plant or animal and wondered just how the thing worked were in the presence of unfathomable technology. They were truly in the dark.

The earliest biological investigations began in the only way they could—with the naked eye.[2] A number of books from about 400 B.C. (attributed to Hippocrates, the "father of medicine") describe the symptoms of some common diseases and attribute sickness to diet and other physical causes, rather than to the work of the gods. Although the writings were a beginning, the ancients were still lost when it came to the composition of living things. They believed that all matter was made up of four elements: earth, air, fire, and water. Living bodies were thought to be made of four "humors"—blood, yellow bile, black bile, and phlegm—and all disease supposedly arose from an excess of one of the humors.

The greatest biologist of the Greeks was also their greatest philosopher, Aristotle. Born when Hippocrates was still alive, Aristotle realized (unlike almost everyone before him) that knowledge of nature requires systematic observation. Through careful examination he recognized an astounding amount of order within the living world, a crucial first step. Aristotle grouped animals into two general categories—those with blood, and those without—that correspond closely to the modern classifications of vertebrate and invertebrate. Within the vertebrates he recognized the categories of mammals, birds, and fish. He put most amphibians and reptiles in a single group, and snakes in a separate class. Even though his observations were unaided by instruments, much of Aristotle's reasoning remains sound despite the knowledge gained in the thousands of years since he died.

Only a few significant biological investigators lived during the millennium following Aristotle. One of them was Galen, a second-century A.D. physician in Rome. Galen's work shows that careful observation of the outside and (with dissection) the inside of plants and animals, although necessary, is not sufficient to comprehend biology. For example, Galen tried to understand the function of animal organs. Although he knew that the heart pumped blood, he could not tell just from looking that the blood circulated and returned to the heart.

Galen mistakenly thought that blood was pumped out to "irrigate" the tissues, and that new blood was made continuously to resupply the heart. His idea was taught for nearly fifteen hundred years.

It was not until the seventeenth century that an Englishman, William Harvey, introduced the theory that blood flows continuously in one direction, making a complete circuit and returning to the heart. Harvey calculated that if the heart pumps out just two ounces of blood per beat, at 72 beats per minute, in one hour it would have pumped 540 pounds of blood—triple the weight of a man! Since making that much blood in so short a time is clearly impossible, the blood had to be reused. Harvey's logical reasoning (aided by the still-new Arabic numerals, which made calculating easy) in support of an unobservable activity was unprecedented; it set the stage for modern biological thought.

In the Middle Ages the pace of scientific investigation quickened. The example set by Aristotle had been followed by increasing numbers of naturalists. Many plants were described by the early botanists Brunfels, Bock, Fuchs, and Valerius Cordus. Scientific illustration developed as Rondelet drew animal life in detail. The encyclopedists, such as Conrad Gesner, published large volumes summarizing all of biological knowledge. Linnaeus greatly extended Aristotle's work of classification, inventing the categories of class, order, genus, and species. Studies of comparative biology showed many similarities between diverse branches of life, and the idea of common descent began to be discussed.

Biology advanced rapidly in the seventeenth and eighteenth centuries as scientists combined Aristotle's and Harvey's examples of attentive observation and clever reasoning. Yet even the strictest attention and cleverest reasoning will take you only so far if important parts of a system aren't visible. Although the human eye can resolve objects as small as one-tenth of a millimeter, a lot of the action in life occurs on a micro level, a Lilliputian scale. So biology reached a plateau: One black box, the gross structure of organisms, was opened only to reveal the black box of the finer levels of life. In order to proceed further biology needed a series of technological breakthroughs. The first was the microscope.

BLACK BOXES WITHIN BLACK BOXES

Lenses were known in ancient times, and by the fifteenth century their use in spectacles was common. It was not until the seventeenth century, however, that a convex and a concave lens were put together in a tube to form the first crude microscope. Galileo used one of the first instruments, and he was amazed to discover the compound eyes of insects. Stelluti looked at the eyes, tongue, antennae, and other parts of bees and weevils. Malpighi confirmed the circulation of the blood through capillaries and he described the early development of the embryonic chick heart. Nehemiah Grew inspected plants; Swammerdam dissected the mayfly; Leeuwenhoek was the first person ever to see a bacterial cell; and Robert Hooke described cells in cork and leaves (although their importance escaped him.)

The discovery of an unanticipated Lilliputian world had begun, overturning settled notions of what living things are. Charles Singer, the historian of science, noted that "the infinite complexity of living things thus revealed was as philosophically disturbing as the ordered majesty of the astronomical world which Galileo had unveiled to the previous generation, though it took far longer for its implications to sink into men's minds." In other words, sometimes the new boxes demand that we revise all of our theories. In such cases, great unwillingness can arise.

The cell theory of life was finally put forward in the early nineteenth century by Matthias Schleiden and Theodor Schwann. Schleiden worked primarily with plant tissue; he argued for the central importance of a dark spot—the nucleus—within all cells. Schwann concentrated on animal tissue, in which it was harder to see cells. Nonetheless he discerned that animals were similar to plants in their cellular structure. Schwann concluded that cells or the secretions of cells compose the entire bodies of animals and plants, and that in some way the cells are individual units with a life of their own. He wrote that "the question as to the fundamental power of organized bodies resolves itself into that of individual cells." As Schleiden added, "Thus the primary question is, what is the origin of this peculiar little organism, the cell?"

Schleiden and Schwann worked in the early to middle 1800s—the time of Darwin's travels and the writing of The Origin of Species. To Darwin, then, as to every other scientist of the time, the cell was a black box. Nonetheless he was able to make sense of much biology above the level of the cell. The idea that life evolves was not original with Darwin, but he argued it by far the most systematically, and the theory of how evolution works—by natural selection working on variation—was his own.

Meanwhile, the cellular black box was steadily explored. The investigation of the cell pushed the microscope to its limits, which are set by the wavelength of light. For physical reasons a microscope cannot resolve two points that are closer together than approximately one-half of the wavelength of the light that is illuminating them. Since the wavelength of visible light is roughly one-tenth the diameter of a bacterial cell, many small, critical details of cell structure simply cannot be seen with a light microscope. The black box of the cell could not be opened without further technological improvements.

In the late nineteenth century, as physics progressed rapidly, J. J. Thomson discovered the electron; the invention of the electron microscope followed several decades later. Because the wavelength of the electron is shorter than the wavelength of visible light, much smaller objects can be resolved if they are "illuminated" with electrons. Electron microscopy has a number of practical difficulties, not least of which is the tendency of the electron beam to fry the sample. But ways were found to get around the problems, and after World War II electron microscopy came into its own. New subcellular structures were discovered: Holes were seen in the nucleus, and double membranes detected around mitochondria (a cell's power plants). The same cell that looked so simple under a light microscope now looked much different. The same wonder that the early light microscopists felt when they saw the detailed structure of insects was again felt by twentieth-century scientists when they saw the complexities of the cell.

This level of discovery began to allow biologists to approach the greatest black box of all. The question of *how life works* was not one that Darwin or his contemporaries could answer. They knew that eyes were for seeing—but how, exactly, do they see? How does blood clot? How does the body fight disease? The complex structures revealed by the electron microscope were themselves made of smaller components. What were those components? What did they look like? How did they work? The answers to these questions take us out of the realm of biology and into chemistry. They also take us back into the nineteenth century.

THE CHEMISTRY OF LIFE

As anyone can readily see, living things look different from nonliving things. They act different. They feel different, too: Hide and hair can be distinguished easily from rocks and sand. Most people up until the nineteenth century quite naturally thought that life was made of a special kind of material, one different from the material that composed inanimate objects. But in 1828 Friedrich Wöhler heated ammonium cyanate and was amazed to find that urea, a biological waste product, was formed. The synthesis of urea from nonliving material shattered the easy distinction between life and nonlife, and the inorganic chemist Justus von Liebig then began to study the chemistry of life (or biochemistry). Liebig showed that the body heat of animals is due to the combustion of food; it is not simply an innate property of life. From his successes he formulated the idea of metabolism, whereby the body builds up and breaks down substances through chemical processes. Ernst Hoppe-Seyler crystallized the red material from blood (hemoglobin) and showed that it attaches to oxygen in order to carry the latter throughout the body. Emil Fischer demonstrated that the large class of substances called proteins all were constituted from only twenty types of building blocks (called amino acids) joined into chains.

What do proteins look like? Although Emil Fischer showed that they were made of amino acids, the details of their structures were unknown. Their size put them below the reach of even electron microscopy, yet it was becoming clear that proteins were the fundamental machines of life, catalyzing the chemistry and building the structures of the cell. A new technique therefore was needed to study protein structure.

During the first part of the twentieth century, X-ray crystallography was used to determine the structures of small molecules. Crystallography involves shining a beam of X-rays onto a crystal of a chemical; the rays scatter by a process called diffraction. If photographic film is placed behind the crystal, then the diffracted X-rays can be detected by examining the exposed film. The pattern of diffraction can, after the application of strenuous mathematics, indicate the position of *each and every atom* in the molecule. Turning the guns of X-ray crystallography onto proteins would show their structure, but there was a big problem: the more atoms in a molecule, the harder the mathematics, and the harder the task of crystallizing the chemical in the first place. Because proteins have dozens of times more atoms than the molecules typically examined by crystallography, that makes the problem dozens of times more difficult. But some people have dozens of times more perseverance than the rest of us.

In 1958, after decades of work, J. C. Kendrew determined the structure of the protein myoglobin using X-ray crystallography; finally, a technique showed the detailed structure of one of the basic components of life. And what was seen? Once again, more complexity. Before the determination of myoglobin's structure, it was thought that proteins would turn out to be simple and regular structures, like salt crystals. Upon observing the convoluted, complicated, bowel-like structure of myoglobin, however, Max Perutz groaned, "Could the search for ultimate truth really have revealed so hideous and visceral-looking an object?" Biochemists have since grown to like the intricacies of protein structure. Improvements in computers and other instruments make crystallography a lot easier today than it was for Kendrew, although it still requires substantial effort.

As the result of the X-ray work of Kendrew on proteins and (most famously) Watson and Crick on DNA, for the first time biochemists actually knew the shapes of the molecules that they were working on. The beginning of modern biochemistry, which has progressed at a breakneck pace since, can be dated to that time. Advances in physics and chemistry, too, have spilled over and created a strong synergism for research on life.

Although in theory X-ray crystallography can determine the structure of all the molecules of living things, practical problems limit its use to a relatively small number of proteins and nucleic acids. New techniques, though, have been introduced at a dizzying pace to complement and supplement crystallography. One important technique for determining structure is called *nuclear magnetic resonance* (NMR). With NMR a molecule can be studied while in solution—it doesn't have to be tediously crystallized. Like X-ray crystallography, NMR can determine the exact structure of proteins and nucleic acids. Also like crystallography, NMR has limitations that make it usable only with a portion of known proteins. But together NMR and X-ray crystallography have been able to solve the structures of enough proteins to give scientists a detailed understanding of what they look like.

When Leeuwenhoek used a microscope to see a tinier mite on a tiny flea, it inspired Jonathan Swift to write a ditty anticipating an end-less procession of smaller and smaller bugs:

So naturalists observe, a flea
Has smaller fleas that on him prey;
And these have smaller still to bite 'em;
And so proceed *ad infinitum.*

Swift was wrong; the procession does not go on forever. In the late twentieth century we are in the flood tide of research on life, and the end is in sight. The last remaining black box was the cell, which was opened to reveal molecules—the bedrock of nature. Lower we cannot go. Moreover, the work that has already been done on enzymes, other proteins, and nucleic acids has illuminated the principles at work at the ground level of life. Many details remain to be filled in, and some sur-prises undoubtedly remain. But unlike earlier scientists, who looked at a fish or a heart or a cell and wondered what it was and what made it work, modern scientists are satisfied that the actions of proteins and other molecules are sufficient explanations for the basis of life. From Aristotle to modern biochemistry, one layer after another has been peeled away until the cell—Darwin's black box—stands open.

LITTLE JUMPS, BIG JUMPS

Suppose a 4-foot-wide ditch in your backyard, running to the horizon in both directions, separates your property from that of your neigh-bor's. If one day you met him in your yard and asked how he got there, you would have no reason to doubt the answer, "I jumped over the ditch." If the ditch were 8 feet wide and he gave the same answer, you would be impressed with his athletic ability. If the ditch were 15 feet wide, you might become suspicious and ask him to jump again while you watched; if he declined, pleading a sprained knee, you would harbor your doubts but wouldn't be certain that he was just telling a tale. If the "ditch" were actually a canyon 100 feet wide, however, you would not entertain for a moment the bald assertion that he jumped across.

But suppose your neighbor—a clever man—qualifies his claim. He did not come across in one jump. Rather, he says, in the canyon there were a number of buttes, no more than 10 feet apart from one another; he jumped from one narrowly spaced butte to another to reach your side. Glancing toward the canyon, you tell your neighbor that you see no buttes, just a wide chasm separating your yard from his. He agrees, but explains that it took him years and years to come over. During that time buttes occasionally arose in the chasm, and he progressed as they popped up. After he left a butte it usually eroded pretty quickly and crumbled back into the canyon. Very dubious, but with no easy way to prove him wrong, you change the subject to baseball.

This little story teaches several lessons. First, the word *jump* can be offered as an explanation of how someone crossed a barrier, but the explanation can range from completely convincing to totally inadequate depending on details (such as how wide the barrier is). Second, long journeys can be made much more plausible if they are explained as a series of smaller jumps rather than one great leap. And third, in the absence of evidence of such smaller jumps, it is very difficult to prove right or wrong someone who asserts that stepping stones existed in the past but have disappeared.

Of course, the allegory of jumps across narrow ditches versus canyons can be applied to evolution. The word *evolution* has been invoked to explain tiny changes in organisms as well as huge changes. These are often given separate names: Roughly speaking, *microevolution* describes changes that can be made in one or a few small jumps, whereas *macroevolution* describes changes that appear to require large jumps.

The proposal by Darwin that even relatively tiny changes could occur in nature was a great conceptual advance; the observation of such changes was a hugely gratifying confirmation of his intuition. Darwin saw similar but not identical species of finches on the various Galapagos Islands and theorized that they descended from a common ancestor. Recently some scientists from Princeton actually observed the average beak size of finch populations changing over the course of a few years.[3] Earlier it was shown that the numbers of dark- versus light-colored moths in a population changed as the environment went from sooty to clean. Similarly, birds introduced into North America by European settlers have diversified into several distinct groups. In recent decades it has been possible to gain evidence for microevolution on a molecular scale. For instance, viruses such as the one that causes AIDS mutate their coats in order to evade the human immune system. Disease-causing bacteria have made a comeback as strains evolved the ability to defend against antibiotics. Many other examples could be cited.

On a small scale, Darwin's theory has triumphed; it is now about as controversial as an athlete's assertion that he or she could jump over a four-foot ditch. But it is at the level of macroevolution—of large jumps—that the theory evokes skepticism. Many people have followed Darwin in proposing that huge changes can be broken down into plausible, small steps over great periods of time. Persuasive evidence to support that position, however, has not been forthcoming. Nonetheless, like a neighbor's story about vanishing buttes, it has been difficult to evaluate whether the elusive and ill-defined small steps could exist . . . until now.

With the advent of modern biochemistry we are now able to look at the rock-bottom level of life. We can now make an informed evaluation of whether the putative small steps required to produce large evolutionary changes can ever get small enough. You will see in this book that the canyons separating everyday life forms have their counterparts in the canyons that separate biological systems on a microscopic scale. Like a fractal pattern in mathematics, where a motif is repeated even as you look at smaller and smaller scales, unbridgeable chasms occur even at the tiniest level of life.

A SERIES OF EYES

Biochemistry has pushed Darwin's theory to the limit. It has done so by opening the ultimate black box, the cell, thereby making possible our understanding of how life works. It is the astonishing complexity of subcellular organic structures that has forced the question, How could all this have evolved? To feel the brunt of the question—and to get a taste of what's in store for us—let's look at an example of a biochemical system. An explanation for the origin of a function must keep pace with contemporary science. Let's see how science's explanation for one function, vision, has progressed since the nineteenth century, then ask how that affects our task of explaining its origin.

In the nineteenth century, the anatomy of the eye was known in detail. The pupil of the eye, scientists knew, acts as a shutter to let in enough light to see in either brilliant sunlight or nighttime darkness. The lens of the eye gathers light and focuses it on the retina to form a sharp image. The muscles of the eye allow it to move quickly. Different colors of light, with different wavelengths, would cause a blurred image, except that the lens of the eye changes density over its surface to correct for chromatic aberration. These sophisticated methods astounded everyone who was familiar with them. Scientists of the nineteenth century knew that if a person lacked any of the eye's many integrated features, the result would be a severe loss of vision or outright blindness. They concluded that the eye could function only if it were nearly intact.

Charles Darwin knew about the eye, too. In *The Origin of Species* Darwin dealt with many objections to his theory of evolution by natural selection. He discussed the problem of the eye in a section of the book appropriately entitled "Organs of Extreme Perfection and Complication." In Darwin's thinking, evolution could not build a complex organ in one step or a few steps; radical innovations such as the eye would require generations of organisms to slowly accumulate beneficial changes in a gradual process. He realized that if in one generation an organ as complex as the eye suddenly appeared, it would be tantamount to a miracle. Unfortunately, gradual development of the human eye appeared to be impossible, since its many sophisticated features seemed to be interdependent. Somehow, for evolution to be believable, Darwin had to convince the public that complex organs could be formed in a step-by-step process.

He succeeded brilliantly. Cleverly, Darwin didn't try to discover a real pathway that evolution might have used to make the eye. Rather, he pointed to modern animals with different kinds of eyes (ranging from the simple to the complex) and suggested that the evolution of the human eye might have involved similar organs as intermediates (Figure 1–1).

Here is a paraphrase of Darwin's argument: Although humans have complex camera-type eyes, many animals get by with less. Some tiny creatures have just a simple group of pigmented cells—not much more than a light-sensitive spot. That simple arrangement can hardly be said to confer vision, but it can sense light and dark, and so it meets the creature's needs. The light-sensing organ of some starfishes is somewhat more sophisticated. Their eye is located in a depressed region. Since the curvature of the depression blocks off light from some directions, the animal can sense which direction the light is coming from. The directional sense of the eye improves if the curvature becomes more pronounced, but more curvature also lessens the amount of light that enters the eye, decreasing its sensitivity. The sensitivity can be increased by placement of gelatinous material in the cavity to act as a lens; some modern animals have eyes with such

FIGURE 1–1

A SERIES OF EYES. (LEFT) A SIMPLE PATCH OF PHOTORECEPTORS, SUCH AS MAY BE FOUND IN JELLYFISH. (RIGHT) A CUPPED EYE FOUND IN MARINE LIMPETS. (BOTTOM) AN EYE WITH A LENS, FROM A MARINE SNAIL.

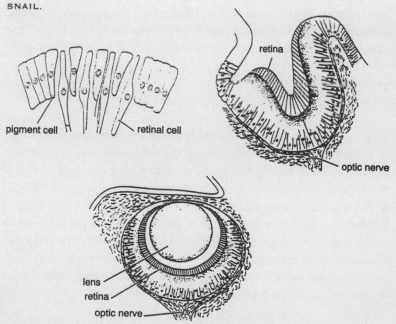

crude lenses. Gradual improvements in the lens could then provide increasingly sharp images to meet the requirements of the animal's environment.

Using reasoning like this, Darwin convinced many of his readers that an evolutionary pathway leads from the simplest light-sensitive spot to the sophisticated camera-eye of man. But the question of how vision began remained unanswered. Darwin persuaded much of the world that a modern eye evolved gradually from a simpler structure, but he did not even try to explain where his starting point—the relatively simple light-sensitive spot—came from. On the contrary, Darwin dismissed the question of the eye's ultimate origin: "How a nerve comes to be sensitive to light hardly concerns us more than how life itself originated."[4]

He had an excellent reason for declining the question: it was completely beyond nineteenth-century science. How the eye works—that is, what happens when a photon of light first hits the retina—simply could not be answered at that time. As a matter of fact, no question about the underlying mechanisms of life could be answered. How did animal muscles cause movement? How did photosynthesis work? How was energy extracted from food? How did the body fight infection? No one knew.

THE VISION OF BIOCHEMISTRY

To Darwin, vision was a black box, but after the cumulative hard work of many biochemists, we are now approaching answers to the question of sight.[5] The following five paragraphs give a biochemical sketch of the eye's operation. (Note: These technical paragraphs are set off by ❐ at the beginning and end.) Don't be put off by the strange names of the components. They're just labels, no more esoteric than *carburetor* or *differential* are to someone reading a car manual for the first time. Readers with an appetite for detail can find more information in many biochemistry textbooks; others may wish to tread lightly, and/or refer to Figures 1–2 and 1–3 for the gist.

❐ When light first strikes the retina a photon interacts with a molecule called 11-*cis*-retinal, which rearranges within picoseconds to *trans*-retinal. (A picosecond is about the time it takes light to

FIGURE 1–2

THE FIRST STEP IN VISION. A PHOTON OF LIGHT CAUSES A CHANGE IN
THE SHAPE OF A SMALL ORGANIC MOLECULE, RETINAL. THIS FORCES A
CHANGE IN THE SHAPE OF THE MUCH LARGER PROTEIN, RHODOPSIN,
TO WHICH IT IS ATTACHED. THE CARTOON DRAWING OF THE PROTEIN IS
NOT TO SCALE.

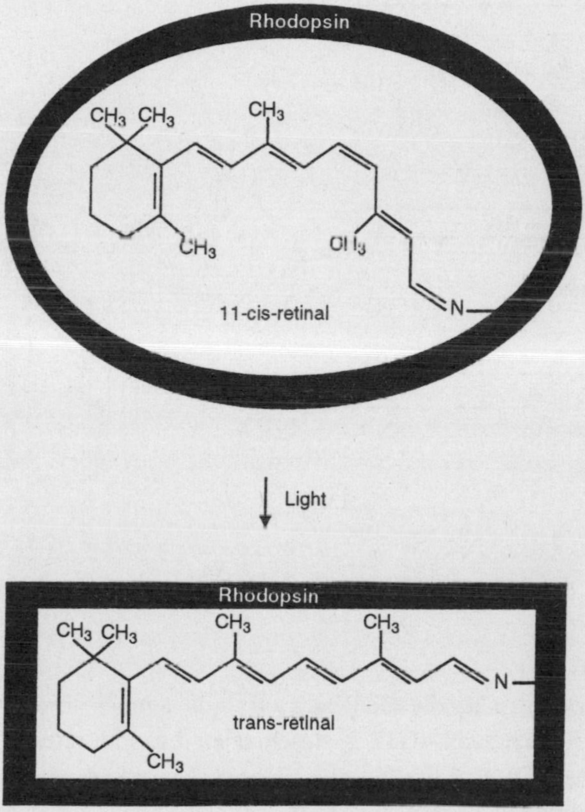

travel the breadth of a single human hair.) The change in the shape
of the retinal molecule forces a change in the shape of the protein,
rhodopsin, to which the retinal is tightly bound. The protein's
metamorphosis alters its behavior. Now called metarhodopsin II,
the protein sticks to another protein, called transducin. Before
bumping into metarhodopsin II, transducin had tightly bound a
small molecule called GDP. But when transducin interacts with

FIGURE 1–3

THE BIOCHEMISTRY OF VISION. RH, RHODOPSIN; RHK, RHODOPSIN
KINASE; A, ARRESTIN; GC, GUANYLATE CYCLASE; T, TRANSDUCIN;
PDE, PHOSPHODIESTERASE.

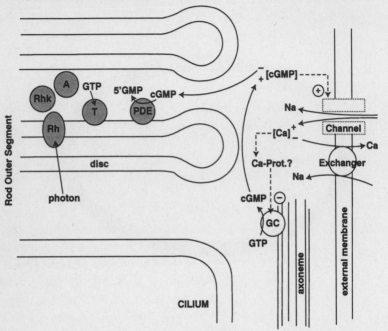

From Chabre, M. & Deterre, P. (1989) *European Journal of Biochemistry, 179,* 255. Reproduced with permission.

metarhodopsin II, the GDP falls off, and a molecule called GTP
binds to transducin. (GTP is closely related to, but critically different from, GDP.)

GTP-transducin-metarhodopsin II now binds to a protein called
phosphodiesterase, located in the inner membrane of the cell.
When attached to metarhodopsin II and its entourage, the phosphodiesterase acquires the chemical ability to "cut" a molecule
called cGMP (a chemical relative of both GDP and GTP). Initially
there are a lot of cGMP molecules in the cell, but the phosphodiesterase lowers its concentration, just as a pulled plug lowers the
water level in a bathtub.

Another membrane protein that binds cGMP is called an ion channel. It acts as a gateway that regulates the number of sodium ions in the cell. Normally the ion channel allows sodium ions to flow into the cell, while a separate protein actively pumps them out again. The dual action of the ion channel and pump keeps the level of sodium ions in the cell within a narrow range. When the amount of cGMP is reduced because of cleavage by the phosphodiesterase, the ion channel closes, causing the cellular concentration of positively charged sodium ions to be reduced. This causes an imbalance of charge across the cell membrane that, finally, causes a current to be transmitted down the optic nerve to the brain. The result, when interpreted by the brain, is vision.

If the reactions mentioned above were the only ones that operated in the cell, the supply of 11-cis-retinal, cGMP, and sodium ions would quickly be depleted. Something has to turn off the proteins that were turned on and restore the cell to its original state. Several mechanisms do this. First, in the dark the ion channel (in addition to sodium ions) also lets calcium ions into the cell. The calcium is pumped back out by a different protein so that a constant calcium concentration is maintained. When cGMP levels fall, shutting down the ion channel, calcium ion concentration decreases, too. The phosphodiesterase enzyme, which destroys cGMP, slows down at lower calcium concentration. Second, a protein called guanylate cyclase begins to resynthesize cGMP when calcium levels start to fall. Third, while all of this is going on, metarhodopsin II is chemically modified by an enzyme called rhodopsin kinase. The modified rhodopsin then binds to a protein known as arrestin, which prevents the rhodopsin from activating more transducin. So the cell contains mechanisms to limit the amplified signal started by a single photon.

Trans-retinal eventually falls off of rhodopsin and must be reconverted to 11-cis-retinal and again bound by rhodopsin to get back to the starting point for another visual cycle. To accomplish this, trans-retinal is first chemically modified by an enzyme to trans-retinol—a form containing two more hydrogen atoms. A second enzyme then converts the molecule to 11-cis-retinol. Finally, a third enzyme removes the previously added hydrogen atoms to form 11-cis-retinal, a cycle is complete. ❐

The above explanation is just a sketchy overview of the biochemistry of vision. Ultimately, though, *this* is the level of explanation for which biological science must aim. In order to truly understand a function, one must understand in detail every relevant step in the process. The relevant steps in biological processes occur ultimately at the molecular level, so a satisfactory explanation of a biological phenomenon—such as sight, digestion, or immunity—must include its molecular explanation.

Now that the black box of vision has been opened, it is no longer enough for an evolutionary explanation of that power to consider only the *anatomical* structures of whole eyes, as Darwin did in the nineteenth century (and as popularizers of evolution continue to do today). Each of the anatomical steps and structures that Darwin thought were so simple actually involves staggeringly complicated biochemical processes that cannot be papered over with rhetoric. Darwin's metaphorical hops from butte to butte are now revealed in many cases to be huge leaps between carefully tailored machines—distances that would require a helicopter to cross in one trip.

Thus biochemistry offers a Lilliputian challenge to Darwin. Anatomy is, quite simply, irrelevant to the question of whether evolution could take place on the molecular level. So is the fossil record. It no longer matters whether there are huge gaps in the fossil record or whether the record is as continuous as that of U.S. presidents. And if there are gaps, it does not matter whether they can be explained plausibly.[6] The fossil record has nothing to tell us about whether the interactions of 11-*cis*-retinal with rhodopsin, transducin, and phosphodiesterase could have developed step-by-step. Neither do the patterns of biogeography matter, nor those of population biology, nor the traditional explanations of evolutionary theory for rudimentary organs or species abundance. This is not to say that random mutation is a myth, or that Darwinism fails to explain anything (it explains microevolution very nicely), or that large-scale phenomena like population genetics don't matter. They do. Until recently, however, evolutionary biologists could be unconcerned with the molecular details of life because so little was known about them. Now the black box of the cell has been opened, and the infinitesmal world that stands revealed must be explained.

CALVINISM

It seems to be characteristic of the human mind that when it sees a black box in action, it imagines that the contents of the box are simple. A happy example is seen in the comic strip "Calvin and Hobbes" (Figure 1–4). Calvin is always jumping in a box with his stuffed tiger, Hobbes, and traveling back in time, or "transmogrifying" himself into animal shapes, or using it as a "duplicator" and making clones of himself. A little boy like Calvin easily imagines that a box can fly like an airplane (or something), because Calvin doesn't know how airplanes work.

In some ways, grown-up scientists are just as prone to wishful thinking as little boys like Calvin. For example, centuries ago it was thought that insects and other small animals arose directly from spoiled food. This was easy to believe, because small animals were thought to be very simple (before the invention of the microscope, naturalists thought that insects had no internal organs.) But as biology progressed and careful experiments showed that protected food did not breed life, the theory of spontaneous generation retreated to the limits beyond which science could not detect what was really happening. In the nineteenth century that meant the cell. When beer, milk, or urine were allowed to sit for several days in containers, even closed ones, they always became cloudy from something growing in them.

FIGURE 1–4
CALVIN AND HOBBES FLY IN THEIR BLACK BOX.

Calvin and Hobbes   by Bill Watterson

The microscopes of the eighteenth and nineteenth centuries showed that the growth was very small, apparently living cells. So it seemed reasonable that simple living organisms could arise spontaneously from liquids.

The key to persuading people was the portrayal of the cells as "simple." One of the chief advocates of the theory of spontaneous generation during the middle of the nineteenth century was Ernst Haeckel, a great admirer of Darwin and an eager popularizer of Darwin's theory. From the limited view of cells that microscopes provided, Haeckel believed that a cell was a "simple little lump of albuminous combination of carbon,"[7] not much different from a piece of microscopic Jell-O. So it seemed to Haeckel that such simple life, with no internal organs, could be produced easily from inanimate material. Now, of course, we know better.

Here is a simple analogy: Darwin is to our understanding of the origin of vision as Haeckel is to our understanding of the origin of life. In both cases brilliant nineteenth-century scientists tried to explain Lilliputian biology that was hidden from them, and both did so by assuming that the inside of the black box must be simple. Time has proven them wrong.

In the first half of the twentieth century, the many branches of biology did not often communicate with each other.[8] As a result genetics, systematics, paleontology, comparative anatomy, embryology, and other areas developed their own views of what evolution meant. Inevitably, evolutionary theory began to mean different things to different disciplines; a coherent view of Darwinian evolution was being lost. In the middle of the century, however, leaders of the fields organized a series of interdisciplinary meetings to combine their views into a coherent theory of evolution based on Darwinian principles. The result has been called the "evolutionary synthesis," and the theory called neo-Darwinism. Neo-Darwinism is the basis of modern evolutionary thought.

One branch of science was not invited to the meetings, and for good reason: it did not yet exist. The beginnings of modern biochemistry came only after neo-Darwinism had been officially launched. Thus, just as biology had to be reinterpreted after the complexity of microscopic life was discovered, neo-Darwinism must be reconsidered in light of advances in biochemistry. The scientific disciplines that were part of the evolutionary synthesis are all nonmolecular. Yet for the Darwinian theory of evolution to be true, it has to account for the molecular structure of life. It is the purpose of this book to show that it does not.

CHAPTER 2

NUTS AND BOLTS

THE NATIVES ARE RESTLESS

Lynn Margulis is Distinguished University Professor of Biology at the University of Massachusetts. Lynn Margulis is highly respected for her widely accepted theory that mitochondria, the energy source of plant and animal cells, were once independent bacterial cells. And Lynn Margulis says that history will ultimately judge neo-Darwinism as "a minor twentieth-century religious sect within the sprawling religious persuasion of Anglo-Saxon biology."[1] At one of her many public talks she asks the molecular biologists in the audience to name a single, unambiguous example of the formation of a new species by the accumulation of mutations. Her challenge goes unmet. Proponents of the standard theory, she says, "wallow in their zoological, capitalistic, competitive, cost-benefit interpretation of Darwin—having mistaken him. . . . Neo-Darwinism, which insists on (the slow accrual of mutations), is in a complete funk."

Juicy quotes, these. And she is not alone in her unhappiness. Over the past 130 years Darwinism, although securely entrenched, has met a steady stream of dissent both from within the scientific community and from without it. In the 1940s the geneticist Richard Goldschmidt became so disenchanted with Darwinism's explanation for the origins of new structures that he was driven to propose the "hopeful monster" theory. Goldschmidt thought that occasionally large changes might occur just by chance—perhaps a reptile laid an egg once, say, and from it hatched a bird.

The hopeful-monster theory didn't catch on, but dissatisfaction with a Darwinian interpretation of the fossil record bubbled up several decades later. Paleontologist Niles Eldredge describes the problem:[2]

> No wonder paleontologists shied away from evolution for so long. It never seems to happen. Assiduous collecting up cliff faces yields zigzags, minor oscillations, and the very occasional slight accumulation of change—over millions of years, at a rate too slow to account for all the prodigious change that has occurred in evolutionary history. When we do see the introduction of evolutionary novelty, it usually shows up with a bang, and often with no firm evidence that the fossils did not evolve elsewhere! Evolution cannot forever be going on somewhere else. Yet that's how the fossil record has struck many a forlorn paleontologist looking to learn something about evolution.[2]

To try to alleviate the dilemma, in the early 1970s Eldredge and Stephen Jay Gould proposed a theory they called "punctuated equilibrium."[3] The theory postulates two things: that for long periods most species undergo little observable change; and that, when it does occur, change is rapid and concentrated in small, isolated populations. If this happened, then fossil intermediates would be hard to find, squaring with the spotty fossil record. Like Goldschmidt, Eldredge and Gould believe in common descent but think that a mechanism other than natural selection is needed to explain rapid, large-scale changes.

Gould has been at the forefront of the discussion of another fascinating phenomenon: the "Cambrian explosion." Careful searches show only a smattering of fossils of multicellular creatures in rocks older than about 600 million years. Yet in rocks just a little bit younger is seen a profusion of fossilized animals, with a host of widely differing body plans. Recently the estimated time over which the explosion took place has been revised downward from 50 million years to 10 million years—a blink of the eye in geological terms. The shorter time estimate has forced headline writers to grope for new superlatives, a favorite being the "biological Big Bang." Gould has argued that the rapid rate of appearance of new life forms demands a mechanism other than natural selection for its explanation.[4]

Ironically, we have come full circle from Darwin's day. When Darwin first proposed his theory a big difficulty was the estimated age of the earth. Nineteenth-century physicists thought the earth was only about a hundred million years old, yet Darwin thought natural selection would require much more time to produce life. At first he was proven right; the earth is now known to be much older. With the discovery of the biological Big Bang, however, the window of time for life to go from simple to complex has shrunk to much less than nineteenth-century estimates of the earth's age.

It is not just paleontologists looking for bones, though, who are disgruntled. A raft of evolutionary biologists examining whole organisms wonder just how Darwinism can account for their observations. The English biologists Mae-Wan Ho and Peter Saunders complain as follows:

> It is now approximately half a century since the neo-Darwinian synthesis was formulated. A great deal of research has been carried on within the paradigm it defines. Yet the successes of the theory are limited to the minutiae of evolution, such as the adaptive change in coloration of moths; while it has remarkably little to say on the questions which interest us most, such as how there came to be moths in the first place.[5]

University of Georgia geneticist John McDonald notes a conundrum:

> The results of the last 20 years of research on the genetic basis of adaptation has led us to a great Darwinian paradox. *Those* [genes] *that are obviously variable within natural populations do not seem to lie at the basis of many major adaptive changes, while those* [genes] *that seemingly do constitute the foundation of many, if not most, major adaptive changes apparently are not variable within natural populations.*[6] [Emphasis in original]

Australian evolutionary geneticist George Miklos puzzles over the usefulness of Darwinism:

> What then does this all-encompassing theory of evolution predict? Given a handful of postulates, such as random mutations, and selection coefficients, it will predict changes in [gene] frequencies over time. Is this what a grand theory of evolution ought to be about?[7]

Jerry Coyne, of the Department of Ecology and Evolution at the University of Chicago, arrives at an unanticipated verdict:

> We conclude—unexpectedly—that there is little evidence for the neo-Darwinian view: its theoretical foundations and the experimental evidence supporting it are weak.[8]

And University of California geneticist John Endler ponders how beneficial mutations arise:

> Although much is known about mutation, it is still largely a "black box" relative to evolution. Novel biochemical functions seem to be rare in evolution, and the basis for their origin is virtually unknown.[9]

Mathematicians over the years have complained that Darwinism's numbers just do not add up. Information theorist Hubert Yockey argues that the information needed to begin life could not have developed by chance; he suggests that life be considered a given, like matter or energy.[10] In 1966 leading mathematicians and evolutionary biologists held a symposium at the Wistar Institute in Philadelphia because the organizer, Martin Kaplan, had overheard "a rather weird discussion between four mathematicians . . . on mathematical doubts concerning the Darwinian theory of evolution."[11] At the symposium one side was unhappy, and the other was uncomprehending. A mathematician who claimed that there was insufficient time for the number of mutations apparently needed to make an eye was told by the biologists that his figures must be wrong. The mathematicians, though, were not persuaded that the fault was theirs. As one said:

> There is a considerable gap in the neo-Darwinian theory of evolution, and we believe this gap to be of such a nature that it cannot be bridged with the current conception of biology.[12]

Stuart Kauffman of the Santa Fe Institute is a leading proponent of "complexity theory." Simply put, it proposes that many features of living systems are the result of self-organization—the tendency of complex systems to arrange themselves in patterns—and not natural selection:

> Darwin and evolution stand astride us, whatever the mutterings of creation scientists. But is the view right? Better, is it adequate? I believe it is not. It is not that Darwin is wrong, but that he got hold of only part of the truth.[13]

Complexity theory has so far attracted few followers but much criticism. John Maynard Smith, under whom Kauffman did graduate work, complains that the theory is too mathematical and is unconnected to real-life chemistry.[14] Although the complaint has merit, Smith offers no solution to the problem which Kauffman identified—the origin of complex systems.

All told, Darwin's theory has generated dissent from the time it was published, and not just for theological reasons. In 1871 one of Darwin's critics, St. George Mivart, listed his objections to the theory, many of which are surprisingly similar to those raised by modern critics.

> What is to be brought forward (against Darwinism) may be summed up as follows: That "Natural Selection" is incompetent to account for the incipient stages of useful structures. That it does not harmonize with the co-existence of closely similar structures of diverse origin. That there are grounds for thinking that specific differences may be developed suddenly instead of gradually. That the opinion that species have definite though very different limits to their variability is still tenable. That certain fossil transitional forms are absent, which might have been expected to be present. . . . That there are many remarkable phenomena in organic forms upon which "Natural Selection" throws no light whatever.[15]

It seems, then, that the same argument has gone on without resolution for over a century. From Mivart to Margulis, there have always been well-informed, respected scientists who have found Darwinism to be inadequate. Apparently, either the questions first raised by Mivart have gone unanswered, or some people have not been satisfied by the answers they received.

Before going further we should note the obvious: if a poll were taken of all the scientists in the world, the great majority would say they believed Darwinism to be true. But scientists, like everybody else, base most of their opinions on the word of other people. Of the great majority who accept Darwinism, most (though not all) do so based on authority. Also, and unfortunately, too often criticisms have been dismissed by the scientific community for fear of giving ammunition to creationists. It is ironic that in the name of protecting science, trenchant scientific criticism of natural selection has been brushed aside.

It is time to put the debate squarely in the open, and to disregard public relations problems. The time for the debate is now because at last we have reached the bottom of biology, and a resolution is possible. At the tiniest levels of biology—the chemical life of the cell—we have discovered a complex world that radically changes the grounds on which Darwinian debates must be contested. Consider, for example, what a biochemical view does to the creationist/Darwinist debate about the bombardier beetle.

BEETLE BOMBS

The bombardier beetle is an insect of unassuming appearance, measuring about one half-inch in length. When it is threatened by another bug, however, the beetle has a special method of defending itself, squirting a boiling-hot solution at the enemy out of an aperture in its hind section.[16] The heated liquid scalds its target, which then usually makes other plans for dinner. How is this trick done?

It turns out that the bombardier beetle is using chemistry. Prior to battle, specialized structures called secretory lobes make a very concentrated mixture of two chemicals, hydrogen peroxide and hydroquinone (Figure 2–1). The hydrogen peroxide is the same material as one can buy in a drugstore; hydroquinone is used in photographic development. The mixture is sent into a storage chamber called the collecting vesicle. The collecting vesicle is connected to, but ordinarily sealed off from, a second compartment called (evocatively) the explosion chamber. The two compartments are kept separate from one another by a duct with a sphincter muscle, much like the sphincter muscles upon which humans depend for continence. Attached to the explosion chamber are a number of small knobs called ectodermal glands; these secrete enzyme catalysts into the explosion chamber. When the beetle feels threatened it squeezes muscles surrounding the storage chamber while simultaneously relaxing the sphincter muscle. This forces the solution of hydrogen peroxide and hydroquinone to enter the explosion chamber, where it mixes with the enzyme catalysts.

Now, chemically, things get very interesting. The hydrogen peroxide rapidly decomposes into ordinary water and oxygen, just as a store-bought bottle of hydrogen peroxide will decompose over time if left open. The oxygen reacts with the hydroquinone to yield more water, plus a highly irritating chemical called quinone. These reactions release a large quantity of heat. The temperature of the solution rises to

FIGURE 2-1

DEFENSIVE APPARATUS OF A BOMBARDIER BEETLE: B, COLLECTING
VESICLE; E, EXPLOSION CHAMBER; G, ECTODERMAL GLANDS SECRETING
CATALASE; L, SECRETORY LOBES; M, SPHINCTER MUSCLE; O, OUTLET
DUCT. B CONTAINS A MIXTURE OF HYDROQUINONE AND HYDROGEN
PEROXIDE, EXPLODED BY CATALASE WHEN IT PASSES INTO E.

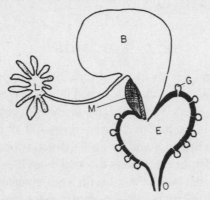

From Crowson, R. A. (1981) *The Biology of the Coleoptera*, Academic Press, New York, chapter 15.
Reproduced with permission.

the boiling point; in fact, a portion vaporizes into steam. The steam
and oxygen gas exert a great deal of pressure on the walls of the explosion chamber. With the sphincter muscle now closed, a channel leading outward from the beetle's body provides the only exit for the boiling mixture. Muscles surrounding the channel allow the steam jet to
be directed at the source of danger. The end result is that the beetle's
enemy is scalded by a steaming solution of the toxic chemical
quinone.

You may wonder why the mixture of hydrogen peroxide and
quinone did not react explosively when they were in the collecting
vesicle. The reason is that many chemical reactions occur quite slowly
if there is no easy way for the molecules to get together on the atomic
level—otherwise, this book would burst into flame as it reacted with
oxygen in the air. As an analogy, consider a locked door. There is no
easy way for people (say, teenage boys and girls) on opposite sides of
the door to get together, even if they would be happy to do so. If someone has the key, however, then the door can be opened and proper introductions can be made. The enzyme catalysts play the role of the

key, allowing the hydrogen peroxide and hydroquinone to get together on the atomic level so that a reaction can take place.

The bombardier beetle is a favorite of creationists. (A storybook for children, *Bomby, the Bombardier Beetle* by Hazel May Rue, has been published by the Institute for Creation Research.) They twit evolutionists with the beetle's remarkable defensive system, inviting them to explain how it could have evolved gradually. Richard Dawkins, professor of zoology at Oxford University, has taken up their challenge. Dawkins is the best modern popularizer of Darwinism around. His books, including the critically acclaimed *The Blind Watchmaker*, are accessible to the interested layman and very entertaining to boot. Dawkins writes with passion because he believes Darwinism is true. He also believes that atheism is a logical deduction from Darwinism and that the world would be better off if more people shared that view.

In *The Blind Watchmaker* Dawkins turns his attention briefly to the bombardier beetle. First he cites a passage from *The Neck of the Giraffe,* a book by science writer Francis Hitching, that describes the bombardier beetle's defensive system, as part of an argument against Darwinism:

> [The bombardier beetle] squirts a lethal mixture of hydroquinone and hydrogen peroxide into the face of its enemy. These two chemicals, when mixed together, literally explode. So in order to store them inside its body, the bombardier beetle has evolved a chemical inhibitor to make them harmless. At the moment the beetle squirts the liquid out of its tail, an anti-inhibitor is added to make the mixture explosive once again. The chain of events that could have led to the evolution of such a complex, co-ordinated and subtle process is beyond biological explanation on a simple step-by-step basis. The slightest alteration in the chemical balance would result immediately in a race of exploded beetles.[17]

Replies Dawkins:

> A biochemist colleague has kindly provided me with a bottle of hydrogen peroxide, and enough hydroquinone for 50 bombardier beetles. I am about to mix the two together. According to [Hitching], they will explode in my face. Here goes. . . . Well, I'm still here. I poured the hydrogen peroxide into the hydroquinone, and absolutely nothing happened. It didn't even get warm. . . . The statement that "these two chemicals, when mixed

together, literally explode," is, quite simply, false, although it is regularly repeated throughout the creationist literature. If you are curious about the bombardier beetle, by the way, what actually happens is as follows. It is true that it squirts a scaldingly hot mixture of hydrogen peroxide and hydroquinone at enemies. But hydrogen peroxide and hydroquinone don't react violently together unless a catalyst is *added*. This is what the bombardier beetle does. As for the evolutionary precursors of the system, both hydrogen peroxide and various kinds of quinones are used for other purposes in body chemistry. The bombardier beetle's ancestors simply pressed into different service chemicals that already happened to be around. That's how evolution works.[18]

Although Dawkins gets the better of the exchange, neither he nor the creationists make their case. Dawkins's explanation for the evolution of the system rests on the fact that the system's elements "happened to be around." Thus evolution might be *possible*. But Dawkins has not explained how hydrogen peroxide and quinones came to be secreted together at very high concentration into one compartment that is connected through a sphinctered tube to a second compartment that contains enzymes necessary for the rapid reaction of the chemicals.

The key question is this: How could complex biochemical systems be gradually produced? The problem with the above "debate" is that both sides are talking past each other. One side gets its facts wrong; the other side merely corrects the facts. But the burden of the Darwinians is to answer two questions: First, what exactly *are* the stages of beetle evolution, in all their complex glory? Second, given these stages, how does Darwinism get us from one to the next?

Dawkins didn't give us any details of how the bombardier beetle's defensive system might have evolved. To point out the problem with his argument, however, let's use what we know of the beetle's anatomy to build the best possible case for the evolution of the bombardier beetle. First, we should note that the function of the bombardier beetle's defensive apparatus is to repel attackers. The components of the system are (1) hydrogen peroxide and hydroquinone, which are produced by the secretory lobes; (2) the enzyme catalysts, which are made by the ectodermal glands; (3) the collecting vesicle; (4) the sphincter muscle; (5) the explosion chamber; and (6) the outlet duct. Not all of these components, though, are necessary for the function of the system. Hydroquinone itself is noxious to predators. A large number of beetle species synthesize quinones that are not even secreted, but which "taste bad." Initially a number of individual beetles are chewed up and spit out, but a predator learns to avoid their noxious counterparts in the future, and thus the species as a whole benefits from this defense.

Hydroquinone alone, then, has the defensive function that we ascribed to the whole system. Can the other components be added to the bombardier's system in such a way that the function continuously improves? It would seem that they can. We can imagine that the beetle would benefit from concentrating the hydroquinone in a holding space such as the collecting vesicle. This would allow the beetle to make a large amount of the noxious chemical, and in so doing become very untasty, without causing internal problems. If the collecting vesicle somehow developed a channel to the outside, the hydroquinone could leak and perhaps repel attackers before they actually ate the bug. Many beetles have defensive apparatuses called pygidial glands that have this basic structure: a simple holding space with a duct leading to the outside, often surrounded by a muscle to help expel the contents of the space. This might be improved by developing a sphincter muscle that would prevent the contents from leaking until the proper time.

Indeed, hydrogen peroxide is also an irritant, and so a beetle might be safer if it could secrete, even at low temperature, both hydroquinone and hydrogen peroxide in order to increase the irritant effect. Almost all cells carry an enzyme called catalase, which breaks down hydrogen peroxide into water and oxygen with the release of heat. If cells lining the tract to the outside secreted a little bit of catalase, then during ejection some of the hydrogen peroxide would be decomposed, warming the solution and thereby making it more irritating. Bombardier beetle species from Australia[19] and Papua New Guinea[20] spray solutions that range in temperature from warm to hot, but not boiling. If the cells released more catalase the solution would become hotter; eventually an optimum would be reached between the hotness of the solution and the durability of the exit channel. Over time the exit channel could be toughened and expanded to allow increased temperature right up to the boiling point of the solution. Subsequent secretion of peroxidases into the catalytic mixture would give an apparatus essentially identical to that pictured in Figure 2–1.

Now we have a scenario fit for the evolutionary literature. But has the development of the defensive apparatus of the bombardier beetle truly been explained? Unfortunately, the explanation here is no more detailed than Darwin's nineteenth-century story about the eye. Although we seem to have a continuously changing system, the components that control its operation are not known. For example, the collection vesicle is a complex, multicelled structure. What does it contain? Why does it have its particular shape? Saying that "the beetle would benefit from concentrating the hydroquinone in a holding space" is like saying "society benefits from concentrating power in a centralized government": In both cases the manner of concentrating and the holding vessel are unexplained, and the benefits of either would depend sharply on the details. The collecting vesicle, the sphincter muscle, the explosion chamber, and the exit port are all complex structures in their own right, with many unidentified components. Furthermore, the actual processes responsible for the development of the explosive capability are unknown: What causes a collection vesicle to develop, hydrogen peroxide to be excreted, or a sphincter muscle to wrap around?

All we can conclude at this point is that Darwinian evolution *might* have occured. If we could analyze the structural details of the beetle down to the last protein and enzyme, and if we could account for all these details with a Darwinian explanation, then we could agree with Dawkins. For now, though, we cannot tell whether the step-by-step accretions of our hypothetical evolutionary stream are single-mutation "hops" or helicopter rides between distant buttes.

SEEING IS BELIEVING

Let's go back to the human eye. Dawkins and Hitching also clash over this classic complex organ. Hitching had stated in *The Neck of the Giraffe* that

it is quite evident that if the slightest thing goes wrong en route—if the cornea is fuzzy, or the pupil fails to dilate, or the lens becomes opaque, or the focusing goes wrong—then a recognizable image is not formed. The eye either functions as a whole or not at all. So how did it come to evolve by slow, steady, infinitesimally small Darwinian improvements? Is it really plausible that thousands upon thousands of lucky chance mutations happened coincidentally so that the lens and the retina, which cannot work without each other, evolved in synchrony? What survival value can there be in an eye that doesn't see?[21]

Dawkins, grateful that Hitching again leads with his chin, doesn't miss the opportunity:

Consider the statement that "if the slightest thing goes wrong . . . [if] the focusing goes wrong . . . a recognizable image is not formed." The odds cannot be far from 50/50 that you are reading these words through glass lenses. Take them off and look around. Would you agree that "a recogniz-able image is not formed"? . . . (Hitching) also states, as though it were ob-vious, that the lens and the retina cannot work without each other. On what authority? Someone close to me has had a cataract operation in both eyes. She has no lenses in her eyes at all. Without glasses she couldn't even begin to play lawn tennis or aim a rifle. But she assures me that you are far better off with a lensless eye than with no eye at all. You can tell if you are about to walk into a wall or another person. If you were a wild creature, you could certainly use your lensless eye to detect the looming shape of a predator, and the direction from which it was approaching.[22]

After attacking Hitching—as well as scientists Richard Goldschmidt and Stephen Jay Gould—for worrying about the eye's complexity, Dawkins goes on to paraphrase Charles Darwin's argument for the plausibility of eye evolution:

Some single-celled animals have a light-sensitive spot with a little pigment behind it. The screen shields it from light coming from one direction, which gives it some "idea" of where the light is coming from. Among many-celled animals . . . the pigment-backed light-sensitive cells are set in a little cup. This gives slightly better direction-finding capability. . . . Now, if you make a cup very deep and turn the sides over, you eventually make a lensless pinhole camera. . . . When you have a cup for an eye, almost any vaguely convex, vaguely transparent or even translucent material over its opening will constitute an improvement, because of its slight lens-like properties. Once such a crude proto-lens is there, there is a continuously graded series of improvements, thickening it and making it more transparent and less distorting, the trend culminating in what we would all recognize as a true lens.[23]

We are invited by Dawkins and Darwin to believe that the evolution of the eye proceeded step-by-step through a series of plausible intermediates in infinitesimal increments. But *are* they infinitesimal? Remember that the "light-sensitive spot" that Dawkins takes as his starting point requires a cascade of factors, including 11-*cis*-retinal and rhodopsin, to function. Dawkins doesn't mention them. And where did the "little cup" come from? A ball of cells—from which the cup must be made—will tend to be rounded unless held in the correct shape by molecular supports. In fact, there are dozens of complex proteins involved in maintaining cell shape, and dozens more that control extracellular structure; in their absence, cells take on the shape of so many soap bubbles. Do these structures represent single-step mutations? Dawkins did not tell us how the apparently simple "cup" shape came to be. And although he reassures us that any "translucent material" would be an improvement (recall that Haeckel mistakenly thought it would be easy to produce cells since they were certainly just "simple lumps"), we are not told how difficult it is to produce a "simple lens." In short, Dawkins's explanation is only addressed to the level of what is called gross anatomy.

Both Hitching and Dawkins have misdirected their focus. The eye, or indeed almost any large biological structure, consists of a number of discrete systems. The function of the retina alone is the perception of light. The function of the lens is to gather light and focus it. If a lens is used with a retina, the working of the retina is improved, but both the retina and lens can work by themselves. Similarly, the muscles that focus the lens or turn the eye function as a contraction apparatus, which can be applied to many different systems. The perception of light by the retina is not dependent on them. Tear ducts and eyelids are also complex systems, but separable from the function of the retina.

Hitching's argument is vulnerable because he mistakes an integrated system of systems for a single system, and Dawkins rightly points out the separability of the components. Dawkins, however, merely adds complex systems to complex systems and calls that an explanation. This can be compared to answering the question "How is a stereo system made?" with the words "By plugging a set of speakers into an amplifier, and adding a CD player, radio receiver, and tape deck." Either Darwinian theory can account for the assembly of the speakers and amplifier, or it can't.

IRREDUCIBLE COMPLEXITY AND THE NATURE OF MUTATION

Darwin knew that his theory of gradual evolution by natural selection carried a heavy burden:

> If it could be demonstrated that any complex organ existed which could not possibly have been formed by numerous, successive, slight modifications, my theory would absolutely break down.[4]

It is safe to say that most of the scientific skepticism about Darwinism in the past century has centered on this requirement. From Mivart's concern over the incipient stages of new structures to Margulis's dismissal of gradual evolution, critics of Darwin have suspected that his criterion of failure had been met. But how can we be confident? What type of biological system could not be formed by "numerous, successive, slight modifications"?

Well, for starters, a system that is irreducibly complex. By *irreducibly complex* I mean a single system composed of several well-matched, interacting parts that contribute to the basic function, wherein the removal of any one of the parts causes the system to effectively cease functioning. An irreducibly complex system cannot be produced directly (that is, by continuously improving the initial function, which continues to work by the same mechanism) by slight, successive modifications of a precursor system, because any precursor to an irreducibly complex system that is missing a part is by definition nonfunctional. An irreducibly complex biological system, if there is such a thing, would be a powerful challenge to Darwinian evolution. Since natural selection can only choose systems that are already working, then if a biological system cannot be produced gradually it would have to arise as an integrated unit, in one fell swoop, for natural selection to have anything to act on.

Even if a system is irreducibly complex (and thus cannot have been produced directly), however, one can not definitively rule out the possibility of an indirect, circuitous route. As the complexity of an interacting system increases, though, the likelihood of such an indirect route drops precipitously. And as the number of unexplained, irreducibly complex biological systems increases, our confidence that Darwin's criterion of failure has been met skyrockets toward the maximum that science allows.

In the abstract, it might be tempting to imagine that irreducible complexity simply requires multiple simultaneous mutations—that evolution might be far chancier than we thought, but still possible. Such an appeal to brute luck can never be refuted. Yet it is an empty argument. One may as well say that the world luckily popped into existence yesterday with all the features it now has. Luck is metaphysical speculation; scientific explanations invoke causes. It is almost universally conceded that such sudden events would be irreconcilable with the gradualism Darwin envisioned. Richard Dawkins explains the problem well:

> Evolution is very possibly not, in actual fact, always gradual. But it must be gradual when it is being used to explain the coming into existence of complicated, apparently designed objects, like eyes. For if it is not gradual in these cases, it ceases to have any explanatory power at all. Without gradualness in these cases, we are back to miracle, which is simply a synonym for the total absence of explanation.[25]

The reason why this is so rests in the nature of mutation.

In biochemistry, a mutation is a change in DNA. To be inherited, the change must occur in the DNA of a reproductive cell. The simplest mutation occurs when a single nucleotide (nucleotides are the "building blocks" of DNA) in a creature's DNA is switched to a different nucleotide. Alternatively, a single nucleotide can be added or left out when the DNA is copied during cell division. Sometimes, though, a whole region of DNA—thousands or millions of nucleotides—is accidentally deleted or duplicated. That counts as a single mutation, too, because it happens at one time, as a single event. Generally a single mutation can, at best, make only a small change in a creature—even if the change impresses us as a big one. For example, there is a well-known mutation called *antennapedia* that scientists can produce in a laboratory fruit fly: the poor mutant creature has legs growing out of its head instead of antennas. Although that strikes us as a big change, it really isn't. The legs on the head are typical fruit-fly legs, only in a different location.

An analogy may be useful here: Consider a step-by-step list of instructions. A mutation is a change in *one* of the lines of instructions. So instead of saying, "Take a 1/4-inch nut," a mutation might say, "Take a 3/8-inch nut." Or instead of "Place the round peg in the round hole," we might get "Place the round peg in the square hole." Or instead of "Attach the seat to the top of the engine," we might get "Attach the seat to the handlebars" (but we could only get this if the nuts and bolts could be attached to the handlebars). What a mutation *cannot* do is change all the instructions in one step—say, to build a fax machine instead of a radio.

Thus, to go back to the bombardier beetle and the human eye, the question is whether the numerous anatomical changes can be accounted for by many small mutations. The frustrating answer is that *we can't tell*. Both the bombardier beetle's defensive apparatus and the vertebrate eye contain so many molecular components (on the order of tens of thousands of different types of molecules) that listing them—and speculating on the mutations that might have produced them—is currently impossible. Too many of the nuts and bolts (and screws, motor parts, handlebars, and so on) are unaccounted for. For us to debate whether Darwinian evolution could produce such large structures is like nineteenth century scientists debating whether cells could arise spontaneously. Such debates are fruitless because not all the components are known.

We should not, however, lose our perspective over this; other ages have been unable to answer many questions that interested them. Furthermore, because we can't yet evaluate the question of eye evolution or beetle evolution does not mean we can't evaluate Darwinism's claims for any biological structure. When we descend from the level of a whole animal (such as a beetle) or whole organ (such as an eye) to the molecular level, then in many cases we *can* make a judgment on evolution because all of the parts of many discrete molecular systems *are* known. In the next five chapters we will meet a number of such systems—and render our judgment.

Now, let's return to the notion of irreducible complexity. At this point in our discussion *irreducible complexity* is just a term whose power resides mostly in its definition. We must ask how we can recognize an irreducibly complex system. Given the nature of mutation, when can we be sure that a biological system is irreducibly complex?

The first step in determining irreducible complexity is to specify both the function of the system and all system components. An irreducibly complex object will be composed of several parts, all of which contribute to the function. To avoid the problems encountered with extremely complex objects (such as eyes, beetles, or other multicellular biological systems) I will begin with a simple mechanical example: the humble mousetrap.

The function of a mousetrap is to immobilize a mouse so that it can't perform such unfriendly acts as chewing through sacks of flour or electrical cords, or leaving little reminders of its presence in unswept corners. The mousetraps that my family uses consist of a number of parts (Figure 2–2): (1) a flat wooden platform to act as a base; (2) a metal hammer, which does the actual job of crushing the little mouse; (3) a spring with extended ends to press against the platform and the hammer when the trap is charged; (4) a sensitive catch that releases when slight pressure is applied, and (5) a metal bar that connects to the catch and holds the hammer back when the trap is charged. (There are also assorted staples to hold the system together.)

The second step in determining if a system is irreducibly complex is to ask if all the components are required for the function. In this example, the answer is clearly yes. Suppose that while reading one evening, you hear the patter of little feet in the pantry, and you go to the utility drawer to get a mousetrap. Unfortunately, due to faulty manufacture, the trap is missing one of the parts listed above. Which part could be missing and still allow you to catch a mouse? If the wooden base were gone, there would be no platform for attaching the other components. If the hammer were gone, the mouse could dance all night on the platform without becoming pinned to the wooden base. If there were no spring, the hammer and platform would jangle loosely, and again the rodent would be unimpeded. If there were no catch or metal holding bar, then the spring would snap the hammer shut as soon as you let go of it; in order to use a trap like that you would have to chase the mouse around while holding the trap open.

FIGURE 2–2

A HOUSEHOLD MOUSETRAP

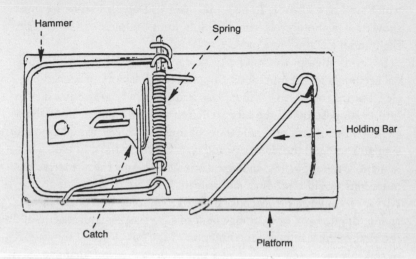

To feel the full force of the conclusion that a system is irreducibly complex and therefore has no functional precursors, we need to distinguish between a *physical* precursor and a *conceptual* precursor. The trap described above is not the only system that can immobilize a mouse. On other occasions my family has used a glue trap. In theory, at least, one can use a box propped open with a stick that could be tripped. Or one can simply shoot the mouse with a BB gun. These are not physical precursors to the standard mousetrap, however, since they cannot be transformed, step by Darwinian step, into a trap with a base, hammer, spring, catch, and holding bar.

To clarify the point, consider this sequence: skateboard, toy wagon, bicycle, motorcycle, automobile, airplane, jet plane, space shuttle. It seems like a natural progression, both because it is a list of objects that all can be used for transportation and also because they are lined up in order of complexity. They can be conceptually connected and blended together into a single continuum. But is, say, a bicycle a physical (and potentially Darwinian) precursor of a motorcycle? No. It is only a *conceptual* precursor. No motorcycle in history, not even the first, was made simply by modifying a bicycle in a stepwise fashion. It might easily be the case that a teenager on a Saturday afternoon could take an old bicycle, an old lawnmower engine, and some spare parts and (with a couple of hours of effort) build himself a functioning motorcycle. But this only shows that humans can design irreducibly complex systems, which we knew already. To be a precursor in Darwin's sense we must show that a motorcycle can be built from "numerous, successive, slight modifications" to a bicycle.

So let us attempt to evolve a bicycle into a motorcycle by the gradual accumulation of mutations. Suppose that a factory produced bicycles, but that occasionally there was a mistake in manufacture. Let us further suppose that if the mistake led to an improvement in the bicycle, then the friends and neighbors of the lucky buyer would demand similar bikes, and the factory would retool to make the mutation a permanent feature. So, like biological mutations, successful mechanical mutations would reproduce and spread. If we are to keep our analogy relevant to biology, however, each change can only be a slight modification, duplication, or rearrangement of a preexisting component, and the change must improve the function of the bicycle. So if the factory mistakenly increased the size of a nut or decreased the diameter of a bolt, or added an extra wheel onto the front axle or left off the rear tire, or put a pedal on the handlebars or added extra spokes, and if any of these slight changes improved the bike ride, then the improvement would immediately be noticed by the buying public and the mutated bikes would, in true Darwinian fashion, dominate the market.

Given these conditions, can we evolve a bicycle into a motorcycle? We can move in the right direction by making the seat more comfortable in small steps, the wheels bigger, and even (assuming our customers prefer the "biker" look) imitating the overall shape in various ways. But a motorcycle depends on a source of fuel, and a bicycle has nothing that can be slightly modified to become a gasoline tank. And what part of the bicycle could be duplicated to begin building a motor? Even if a lucky accident brought a lawnmower engine from a neighboring factory into the bicycle factory, the motor would have to be mounted on the bike and be connected in the right way to the drive chain. How could this be done step-by-step from bicycle parts? A factory that made bicycles simply could not produce a motorcycle by natural selection acting on variation—by "numerous, successive, slight modifications"—and in fact there is no example in history of a complex change in a product occurring in this manner.

A bicycle thus may be a conceptual precursor to a motorcycle, but it is not a physical one. Darwinian evolution requires physical precursors.

MINIMAL FUNCTION

So far we have examined the question of irreducible complexity as a challenge to step-by-step evolution. But there is another difficulty for Darwin. My previous list of factors that render a mousetrap irreducibly complex was actually much too generous, because almost any device with the five components of a standard mousetrap will nonetheless fail to function. If the base were made out of paper, for example, the trap would fall apart. If the hammer were too heavy, it would break the spring. If the spring were too loose, it would not move the hammer. If the holding bar were too short, it would not reach the catch. If the catch were too large, it would not release at the proper time. A simple list of components of a mousetrap is necessary, but not sufficient, to make a functioning mousetrap.

In order to be a candidate for natural selection a system must have *minimal function*: the ability to accomplish a task in physically realistic circumstances. A mousetrap made of unsuitable materials would not meet the criterion of minimal function, but even complex machines that do what they are supposed to do may not be of much use. To illustrate, suppose that the world's first outboard motor had been designed and was being marketed. The motor functioned smoothly— burning gasoline at a controlled rate, transmitting the force along an axle, and turning the propeller—but the propeller rotated at only one revolution per hour. This is an impressive technological feat; after all, burning gasoline in a can next to a propeller doesn't turn it at all. Nonetheless, few people would purchase such a machine, because it fails to perform at a level suitable for its purpose.

Performance can be unsuitable for either of two reasons. The first reason is that the machine does not get the job done. A couple fishing in the middle of a lake in a boat with a slow-turning propeller would not get to the dock: random currents of the water and wind would knock their boat off course. The second reason that performance might be unsuitable is if it is less efficient than can be achieved with simpler means. No one would use an inefficient, outboard motor if they could do just as well or better with a sail.

Unlike irreducible complexity (where we can enumerate discrete parts), minimal function is sometimes hard to define. If one revolution per hour is insufficient for an outboard motor, how about a hundred? Or a thousand? Nonetheless, minimal function is critical in the evolution of biological structures. For example, what is the minimum amount of hydroquinone that a predator can taste? How much of a rise in the temperature of the solution will it notice? If the predator didn't notice a tiny bit of hydroquinone or a small change in temperature, then our Dawkins-esque tale of the bombardier beetle's evolution can be filed alongside the story of the cow jumping over the moon. Irreducibly complex systems are nasty roadblocks for Darwinian evolution; the need for minimal function greatly exacerbates the dilemma.

NUTS AND BOLTS

Biochemistry has demonstrated that any biological apparatus involving more than one cell (such as an organ or a tissue) is necessarily an intricate web of many different, identifiable systems of horrendous complexity. The "simplest" self-sufficient, replicating cell has the capacity to produce thousands of different proteins and other molecules, at different times and under variable conditions. Synthesis, degradation, energy generation, replication, maintenance of cell architecture, mobility, regulation, repair, communication—all of these functions take place in virtually every cell, and each function itself requires the interaction of numerous parts. Because each cell is such an interwoven meshwork of systems, we would be repeating the mistake of Francis Hitching by asking if multicellular structures could have evolved in step-by-step Darwinian fashion. That would be like asking not whether a bicycle could evolve into a motorcycle, but whether a bicycle factory could evolve into a motorcycle factory! Evolution does not take place on the factory level; it takes place on the nut-and-bolt level.

The arguments of Dawkins and Hitching fail because they never discuss what is contained in the systems over which they are arguing. Not only is the eye exceedingly complex, but the "light-sensitive spot" with which Dawkins begins his case is itself a multicelled organ, each of whose cells makes the complexity of a motorcycle or television set look paltry in comparison. Not only does the defensive apparatus of the bombardier beetle depend on a number of interacting components, but the cells that produce hydroquinone and hydrogen peroxide depend on a very large number of components to do so; the cells that secrete catalase are very complex; and the sphincter muscle separating the collection vesicle from the explosion chamber is a system of systems. Because of this, Hitching's arguments about the splendid complexity of the bombardier beetle are easily blurred into irrelevance, and Dawkins's reply satisfies us only until we ask for more details.

In contrast to biological organs, the analysis of simple mechanical objects is relatively straightforward. We showed in short order that a mousetrap is irreducibly complex, and so we can conclude what we already knew—that a mousetrap is made as an intact system. We already knew that a motorcycle was not unconsciously produced by small, successive improvements to a bicycle, and a quick analysis shows us that it is impossible to do so. Mechanical objects can't reproduce and mutate like biological systems, but hypothesizing comparable events at an imaginary factory shows that mutation and reproduction are not the main barriers to evolution of mechanical objects. It is the requirements of the structure–function relationship itself that block Darwinian-style evolution.

Machines are relatively easy to analyze because both their function and all of their parts, each nut and bolt, are known and can be listed. It is then simple to see if any given part is required for the function of the system. If a system requires several closely matched parts to function then it is irreducibly complex, and we can conclude that it was produced as an integrated unit. In principle, biological systems can also be analyzed in this manner, but only if all the parts of the system can be enumerated and a function recognized.

In the past several decades, modern biochemistry has elucidated all or most of the components of a number of biochemical systems. In the next five chapters I will discuss a few of them. In Chapter 3 I will look at a fascinating structure called a "cilium," which some cells use to swim. In the next chapter I will discuss what happens when you cut your finger—and show that the apparent simplicity of blood clotting is deceptively complicated. After that I will consider how cells transport materials from one subcellular compartment to another, encountering many of the same problems that Federal Express meets in delivering packages. In Chapter 6 I will discuss the art of self-defense—on the cellular level, of course. My final biochemical example will be in Chapter 7, where I will look at the intricate system the cell requires just to make one of its "building blocks." In each chapter I will consider whether the system discussed could have developed gradually in a Darwinian fashion, as well as what the scientific community has said about the possible evolution of the systems.

I have endeavored to keep these five "example chapters" as readable and enjoyable as possible. I don't discuss any esoteric concepts peculiar to biochemistry—nothing that is more difficult than the idea of "stick to" or "cut." Nonetheless, as I mentioned in the Preface, to appreciate complexity you have to experience it. The systems I discuss are complex because they contain many components. There is, however, no examination at the end of the book. The detailed descriptions are intended only to give you an appreciation of the complexity of the system, not to test your memory. Some readers may wish to plough right through; others might want to skim and refer back when they are ready for more detail.

I apologize in advance for the complexity of the material, but it is inherent in the point I wish to make. Richard Dawkins can simplify to his heart's content, because he wants to convince his readers that Darwinian evolution is "a breeze." In order to understand the barriers to evolution, however, we have to bite the bullet of complexity.

PART II

EXAMINING THE CONTENTS
OF THE BOX

CHAPTER 3

ROW, ROW, ROW YOUR BOAT

PROTEINS

As strange as it may seem, modern biochemistry has shown that the cell is operated by machines—literally, molecular machines. Like their man-made counterparts (such as mousetraps, bicycles, and space shuttles), molecular machines range from the simple to the enormously complex: mechanical, force-generating machines, like those in muscles; electronic machines, like those in nerves; and solar-powered machines, like those of photosynthesis. Of course, molecular machines are made primarily of proteins, not metal and plastic. In this chapter I will discuss molecular machines that allow cells to swim, and you will see what is required for them to do so.

But first, some necessary details. In order to understand the molecular basis of life one has to have an idea of how proteins work. Those who want to know all the details—how proteins are made, how their structures allow them to work so effectively, and so on—are encouraged to borrow an introductory biochemistry textbook from the library. For those who want to know a few details—such as what amino acids look like, and what are the levels of protein structure—I have included an Appendix that discusses proteins and nucleic acids. For present purposes, however, an overview of these remarkable biochemicals will suffice.

Most people think of proteins as something you eat. In the body of a living animal or plant, however, they play very active roles. Proteins are the machines within living tissue that build the structures and carry out the chemical reactions necessary for life. For example, the first step in capturing the energy in sugar and changing it into a form the body can use is carried out by a catalyzing protein (also known as an enzyme) called hexokinase; skin is made up mostly of a protein called collagen; and when light strikes your retina, the protein called rhodopsin initiates vision. You can see even by this limited number of examples that proteins are amazingly versatile. Nonetheless, a given protein has only one or a few uses: rhodopsin cannot form skin, and collagen cannot interact usefully with light. Therefore a typical cell contains thousands and thousands of different kinds of proteins to perform the many tasks of life.

Proteins are made by chemically hooking together amino acids into a chain. A protein chain typically has anywhere from about fifty to about one thousand amino acid links. Each position in the chain is occupied by one of twenty different amino acids. In this they are like words, which can come in various lengths but are made up from a set of just 26 letters. As a matter of fact, biochemists often refer to each amino acid by a single-letter abbreviation—G for glycine, S for serine, H for histidine, and so forth. Each different kind of amino acid has a different shape and different chemical properties. For example, W is large but A is small, R carries a positive charge but E carries a negative charge, S prefers to be dissolved in water but I prefers oil, and so on.

When you think of a chain, you probably think of something that is very flexible, without much overall shape. But chains of amino acids—in other words, proteins—aren't like that. Proteins that work in a cell fold up into very precise structures, and the structure can be quite different for different types of proteins. The folding is done automatically when, say, a positively charged amino acid attracts a negatively charged one, oil-preferring amino acids huddle together to exclude water, large amino acids are pushed out of small spaces, and so on. Two different amino acid sequences (that is two different proteins) can fold into structures as specific and different from each other as an adjustable wrench and a jigsaw.

It is the shape of a folded protein and the precise positioning of the different kinds of amino acid groups that allow a protein to work (Figure 3–1). For example, if it is the job of one protein to bind specifically to a second protein, then their two shapes must fit each other like a hand in a glove. If there is a positively charged amino acid on the first protein, then the second protein better have a negatively charged amino acid; otherwise, the two will not stick together. If it is the job of a protein to catalyze a chemical reaction, then the shape of the enzyme generally matches the shape of the chemical that is its target. When it binds, the enzyme has amino acids precisely positioned to cause a chemical reaction. If the shape of a wrench or a jigsaw is significantly warped, then the tool doesn't work. Likewise, if the shape of a protein is warped then it fails to do its job.

Modern biochemistry was launched forty years ago when science began to learn what proteins look like. Since then, great strides have been made in understanding exactly how particular proteins carry out particular tasks. In general, the cell's work requires teams of proteins; each member of the team carries out just one part of a larger task. To keep things as simple as possible, in this book I will concentrate on protein teams. Now, let's go swimming.

SWIMMING

Suppose, on a summer day, you find yourself taking a trip to the neighborhood pool for a bit of exercise. After slathering on the sunblock, you lie on a towel reading the latest issue of *Nucleic Acids Research* and wait for the adult swim period to begin. When at long last the whistle blows and the overly energetic younger crowd clears the water, you gingerly dip your toes in. Slowly, painfully, you lower the rest of your body into the surprisingly cold water. Because it would not be dignified, you will not do any cannonballs or fancy dives from the diving board, nor play water volleyball with the younger adults. Rather, you will swim laps.

Pushing off from the side, you bring your right arm up over your head and plunge it into the water, completing one stroke. During the stroke, nerve impulses travel from your brain to your arm muscles, stimulating them to contract in a specific order. The contracting muscles tug against your bones, causing the humerus to rise and rotate.

FIGURE 3–1

(TOP) WHEN TWO PROTEINS BIND SPECIFICALLY, THEIR SHAPES MATCH
EACH OTHER CLOSELY. (BOTTOM) TO CATALYZE A CHEMICAL REACTION,
AN ENZYME POSITIONS GROUPS CLOSE TO THE CHEMICAL THAT IT
BINDS. THE SCISSORS REPRESENTS GROUPS ON THE PROTEIN THAT
WILL CHEMICALLY CUT A SPECIFIC MOLECULE, REPRESENTED BY THE
LIGHT-COLORED SHAPE.

At the same time other muscles squeeze the bones of your fingers together, so that your hand forms a closed cup. Successive nerve impulses provoke other muscles to relax and contract, pulling in various ways on the radius and ulna, and directing the hand downward
into the water. The force of the arm and hand on the water propel
you forward.

After completion of about half of the actions listed above a similar cycle begins, this time with the bones and muscles of the left arm. Simultaneously, nerve impulses travel to the muscles of your legs, causing them to contract and relax rhythmically, pulling the leg bones up and down. Slicing through the water at a stunning two miles per hour, though, you notice that it's getting hard to think; there's a burning sensation in your lungs; and, even though your eyes are open, things start to go black. Ah, yes—you forgot to breathe. It was said of President Ford that he couldn't walk and chew gum at the same time; you find it difficult to coordinate the turning of your head to the water's surface and back again with the other motions required for swimming. Without oxygen to metabolize fuel your brain starts to shut down, preventing conscious nerve impulses from traveling to the distant regions of your body.

Before you pass out and suffer the humiliation of being rescued by a Generation X lifeguard you stop, stand up in the four feet of water, and notice that you're only about twenty feet from the side. To get around the breathing problem, you decide to do the backstroke. The backstroke involves most of the same muscles as freestyle swimming, and allows you to breathe without coordinating neck muscles with everything else. But now you can't see where you're going. Inevitably you drift off course, come too close to the volleyball game, and are smacked in the head by an errant overhand smash.

In order to get far away from the apologetic volleyballers, you decide simply to tread water in the deep end of the pool. Treading water uses your leg muscles, giving you the exercise you want. It also allows both easy breathing and clear vision. After a few minutes, however, your legs begin to cramp. Deep inside your flabby limbs, unknown to you, your seldom-used muscles keep on hand enough fuel for only short bursts of activity, followed by long periods of rest. During the unusually prolonged exercise they quickly run out of sustenance and cease to function effectively. Nerve impulses frantically try to provoke the motions necessary for swimming, but with the muscles malfunctioning, your legs are as useless as a mousetrap with a broken spring.

You relax and remain still. Fortunately, the large region of your body around the waist has a density less than that of water, and so it keeps you afloat. After a minute or two of bobbing in the water, your cramped muscles relax. You spend the rest of the adult swim period floating serenely around the deep end. This doesn't provide much exercise, but at least it is enjoyable—until the whistle blows again, and you are pummeled by the cannonballs of undignified kids.

WHAT IT TAKES

The neighborhood pool scenario illustrates the requirements for swimming. It also shows that efficiency can be improved by adding auxiliary systems to the basic swimming equipment. To take the last scene first, floating requires only that an object be less dense than water; it does not require activity. The ability to float—to be able to keep a portion of the body out of the water with no active effort—can certainly be useful. Yet because the floater simply drifts along with the current, the ability to float is not the same thing as the ability to swim.

A direction-finding system (such as eyesight) is also useful for swimming; however, it is not the same thing as the ability to swim. In the story you could do the backstroke for a while and still advance through the water. Eventually, an inability to sense the surroundings can lead to accidents. Nonetheless, one can swim sighted or one can swim blind.

Swimming clearly requires energy; cramped, useless muscles immediately cause the system to fail. But you traveled twenty feet before running out of oxygen, and then treaded water for a short while before cramping set in. Although they certainly affect the distance a swimmer can go, the size and efficiency of the fuel reserve system thus are not parts of the swimming system itself.

Now let's consider the mechanical requirements of swimming. You used your hands and feet to contact the water and push it, thus moving your body in the opposite direction. Without the limbs, or some substitute, active swimming would be quite impossible. So we can conclude that one requirement for swimming is a paddle. Another requirement is a motor or power source that has at least enough fuel to last several cycles. At the organ level in humans, the motor is the leg or arm muscle that alternately contracts and relaxes. If the muscle is paralyzed, there is no effective motor, and swimming is impossible. The final requirement is for a connection between the motor and the paddling surface: in humans, these are the areas of bones to which the muscles adhere. If a muscle is detached from a bone it can still contract; because it does not move the bone, however, swimming does not take place.

Mechanical examples of swimming systems are easy to find. My youngest daughter has a toy wind-up fish that wiggles its tail, propelling itself somewhat awkwardly through the bathtub. The tail of the toy fish is the paddle surface, the wound spring is the energy source, and a connecting rod transmits the energy. If one of the components—the paddle, motor, or connector—is missing, then the fish goes nowhere. Like a mousetrap without a spring, a swimming system without a paddle, motor, or connector is fatally incomplete. Because the swimming systems need several parts to work, they are irreducibly complex.

Keep in mind that we are discussing only the parts common to all swimming systems—even the most primitive. Additional complexity is frequently seen. For example, my daughter's toy fish has, besides its tail, spring, and connecting rod, several gears that transmit force from the rod to the tail. A propeller-driven ship has all manner of gears and rods redirecting the energy of the motor until it is finally transmitted to the propeller. Unlike the eye of a swimmer, which is separate from the swimming system itself, such extra gears are indeed part of the system—removing them causes the whole setup to grind to a halt. When a real-life system has more than the theoretically minimum number of parts, then you have to check each of the other parts to see if they're required for the system to work.

WHAT ELSE IT TAKES

A simple list of pieces shows the very minimum of requirements. In the last chapter I discussed how a mousetrap that had all the necessary pieces—a hammer, base, spring, catch, and holding bar—still might not work. If the holding bar were too short or the spring too lightweight, for example, the trap would be a failure. Similarly, the pieces of a swimming system must be matched to each other to have at least minimal function. The paddle is necessary, but if its surface is too small a boat might not make enough progress in a required amount of time. Conversely, if the paddle surface is too large, the connector or motor might strain and break when moving. The motor must be strong enough to move the paddle. It must also be regulated to go at an appropriate speed: too slow, and the swimmer does not make physically necessary progress; too fast, and the connector or paddle may break.

But even if we have the right parts of a swimming system, and even if the parts are the right size and strength and are matched to each other, more is needed. The additional requirement—the need to control the timing and direction of the paddle strokes—is easier to see in the example of a human swimmer than in the case of a paddleboat. When a non-swimmer falls into the water he helplessly flails his arms and legs, making no more progress than if he simply floated. Even a beginning swimmer like my oldest daughter, who is just learning the strokes, quickly sinks unless Dad supports her. Her individual strokes are adequate, but their timing is not coordinated, she doesn't hold herself parallel to the water's surface, and she keeps her head out of the water.

Mechanical systems seem not to have those problems. A ship doesn't flail its propeller, and the timing and direction of a paddleboat's strokes are smooth and regular from the beginning. But the argument is deceptive. The apparently effortless abilities are actually built into the shape and connectivity of the paddlewheel, rotor, and motor of the boat. Imagine a steamboat in which the paddle boards were not arranged nicely around a circular frame. Suppose the boards went off at various angles and the rotor turned first forward, then backward, then side to side. Instead of taking a scenic tour of the Mississippi the boat would drift helplessly, spastically floating with the current toward the Gulf of Mexico. A propeller with blades set at haphazard angles would churn water, but it wouldn't move a boat in any particular direction. The apparent ease with which a mechanical system paddles—compared to the difficulties of a human non-swimmer—is an illusion. The engineer who designed the system "trained" it to swim, pushing the water in the correct direction with the correct timing.

In the unforgiving world of nature, an organism spending energy to flail helplessly in the water would have no advantage over the organism floating serenely beside it. Do any cells swim? If so, what swimming systems do they use? Are they, like a Mississippi steamboat, irreducibly complex? Could they have evolved gradually?

THE CILIUM

Some cells swim using a cilium. A cilium is a structure that, crudely put, looks like a hair and beats like a whip. If a cell with a cilium is free to move about in a liquid, the cilium moves the cell much as an oar moves a boat. If the cell is stuck in the middle of a sheet of other cells, the beating cilium moves liquid over the surface of the stationary cell. Nature uses cilia for both jobs. For example, sperm use cilia to swim. In contrast, the stationary cells that line the respiratory tract each have several hundred cilia. The large number of cilia beat in synchrony, much like the oars handled by slaves on a Roman galley ship, to push mucus up to the throat for expulsion. The action removes small foreign particles—like soot—that are accidentally inhaled and stick in the mucus.

Light microscopes showed thin hairs on some cells, but discovery of the Lilliputian details of cilia had to wait for the invention of the electron microscope, which revealed that the cilium is quite a complicated structure. I will be discussing the structure of the cilium for the next few pages. Most readers will probably find the discussion easier to follow by referring frequently to Figure 3–2.

❐ The cilium consists of a membrane-coated bundle of fibers.[1] The ciliary membrane (think of it as a sort of plastic cover) is an outgrowth of the cell membrane, so the interior of the cilium is connected to the interior of the cell. When a cilium is sliced crossways and the cut end is examined by electron microscopy, you see nine rod-like structures around the periphery. The rods are called microtubules. When high-quality photographs are closely inspected, each of the nine microtubules is seen to actually consist of two fused rings. Further examination shows that one of the rings is made from thirteen individual strands. The other ring, joined to the first, is made from ten strands. Summarizing briefly, each of the nine outer microtubules of a cilium is made of a ring of ten strands fused to a ring of thirteen strands.

Biochemical experiments show that microtubules are made from a protein called tubulin. In a cell, tubulin molecules come together like bricks that form a cylindrical smokestack. Each of the nine outer rods is a microtubule that resembles a fused, double-smokestack with bricks of tubulin. Pictures produced by electron mi-

FIGURE 3–2

(TOP) CROSS-SECTION OF A CILIUM SHOWING THE FUSED DOUBLE-RING
STRUCTURE OF THE OUTER MICROTUBULES, THE SINGLE-RING
STRUCTURE OF THE CENTRAL MICROTUBULES, CONNECTING PROTEINS,
AND DYNEIN MOTOR. (BOTTOM) THE SLIDING MOTION INDUCED BY
DYNEIN "WALKING" UP A NEIGHBORING MICROTUBULE IS CONVERTED
TO A BENDING MOTION BY THE FEXIBLE LINKER PROTEIN NEXIN.

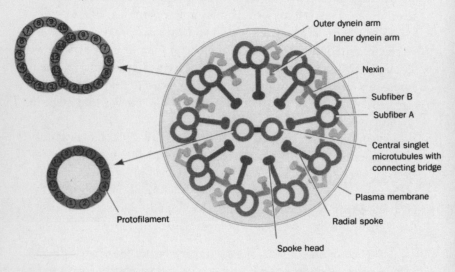

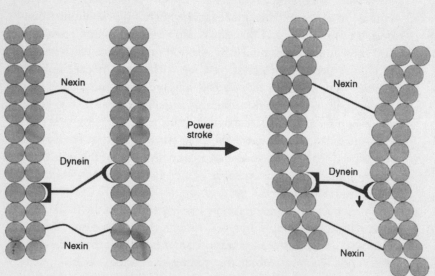

Top from Voet and Voet, fig. 34–77, p. 1256. Reproduced with permission.

croscopy also show two rods in the middle of the cilium. They, too, are microtubules. Instead of being double smokestacks, however, they are individual smokestacks, each made of thirteen strands of tubulin.

When conditions are right within the cell (for example, when the temperature is within certain limits and when the concentration of calcium is just right), tubulin—the "brick" that makes up the smokestacks—automatically comes together to form microtubules. The forces that bring tubulins together are much like those that fold an individual protein into a compact shape: positive charges attract negative charges, oily amino acids squeeze together to exclude water, and so forth. One end of a tubulin molecule has a surface that is complementary to the opposite end of a second tubulin molecule, so the two stick together. A third tubulin can then stick onto the end of the second molecule, a fourth onto the end of the third, and so on. As an analogy, think of the stacking of tuna cans. In the grocery store where my family shops the tuna cans, because the bottom is beveled and is the same diameter as the straight-edged top, stack snugly one on top of the other. If the stack is gently bumped, the cans remain in place.

If two tuna cans are stacked top-to-top instead of top-to-bottom, though, they do not stack securely and can be moved by a casual bump. Furthermore, if Brand X tuna does not have a beveled bottom, it does not stack securely on itself because its cans do not have complementary surfaces. The association of tubulin molecules is much more specific than the stacking of tuna cans. After all, in the cell there are thousands of different proteins, and tubulin has to be sure to associate only with other tubulins—not with just any protein that comes along. Perhaps, then, we should think of tubulin as a tuna can with ten short needle-like projections distributed over the top surface, and ten indentations in the bottom that exactly match the positions of the projections on the top. Now no tuna can will accidentally stack with any other type of can.

Extending our tuna analogy, suppose we also had several projections sticking out one side of the can that were complementary to indentations located almost, but not quite, on the exact opposite side. Then we could stick the cans together side by side and, because the holes were not quite opposite the projections, when we put more cans together they would eventually circle around and form a closed loop. Stacking loops upon loops we eventually (after thoroughly mixing our metaphors) make a structure like a smokestack from our tuna cans.

Although tubulin has the power to self-associate into microtubules, microtubules do not aggregate with one another without help from other proteins. There is a good reason for this: microtubules have a number of jobs to do in the cell. For most jobs, single, unassociated microtubules are needed. For other jobs (including ciliary motion), however, bundles of microtubules are needed. So microtubules lie around individually, like the rods from the game of pick-up sticks, unless purposely bundled together for a particular job.

In photographs of cilia taken by an electron microscope, several different types of connectors can be seen tying together the individual microtubules (see Figure 3–2). There is a protein that bridges the two central single microtubules in the middle of the cilium. Also, from each of the double microtubules, a radial spoke projects toward the center of the cilium. The structure ends in a knobby mass called the spoke head. Finally, a protein called nexin connects each outer, double microtubule to the one beside it.

Two other projections adorn each peripheral microtubule; they are called the outer arm and the inner arm. Biochemical analysis has shown that these projections contain a protein called dynein. Dynein is a member of a class of proteins called motor proteins, which function as tiny motors in the cell, powering mechanical motion. ❐

HOW A CILIUM WORKS

Knowing the structure of a complex machine and knowing how it works are two different matters. One could open the hood of a car and take pictures of the motor until the cows come home, but the snapshots by themselves would not give a clear idea of how the different parts produced the function. Ultimately, in order to find out how a thing works, you have to take it apart and reassemble it, stopping at many points to see if function has yet been restored. Even this may not yield a clear idea of how the machine operates, but it does give a working knowledge of which components are critical. The basic strategy of biochemistry in this century has been to take apart molecular systems and try to put them back together. The strategy has yielded enormous insights into the operations of the cell.

❐ Experiments of this sort have given biochemists clues to how the cilium works. The first clue comes from isolated cilia. Nature has kindly arranged it so that cilia can be separated from cells by vigorous shaking. The shaking breaks off the projections cleanly and, by spinning the solution at high speed (which causes big, heavy particles to sediment more quickly than small, light particles), one can obtain a solution of pure cilia in a test tube. If the cilia are stripped of their membrane and then supplied with a chemical form of energy called ATP, they will beat in characteristic whip-like fashion. This result shows that the motor to power ciliary motion resides in the cilium itself—not in the interior of the now-missing cell. The next clue is that if (through biochemical tricks) the dynein arms are removed but the rest of the cilium is left intact, then the cilium is paralyzed, as if in rigor mortis. Adding back fresh dynein to the stiffened cilia allows motion to resume. So it appears that the motor of the cilium is contained in the dynein arms.

Further experiments gave more clues. There are enzymes (called proteases) that have the ability to chew up other proteins, decomposing them into amino acids. When a small amount of a protease is added for a short time to a solution containing cilia, the protease quickly slices up the nexin linkers at the edge of the structure. The rest of the cilium remains intact. The reason that the protease rapidly attacks the linkers is that, unlike the other proteins of the cilium, the nexin linkers are not folded up tightly; instead, they are loose, flexible chains. Because they are loose, the protease can cut them as rapidly as a pair of scissors can cut a paper ribbon. (The protease cuts tightly folded proteins as rapidly as scissors cut a closed paperback book.)

Proteases allowed biochemists to see how a cilium would work without nexin linkers. What would removal of the linkers do? Perhaps the cilium would work just fine without them, or perhaps it would go into rigor mortis as it did when the dynein arms were removed. In fact, neither of these possibilities occurred. Instead, the linkerless cilium did something quite unexpected. When biochemical energy was supplied to the cilium, instead of bending, it rapidly unraveled. The individual microtubules began to slide past one another like the segments of a radio antenna slide past one another when it is opened. They continued to slide until the length of the cilium had increased by almost tenfold. From this result biochemists concluded that the motor was working, since something had to move the individual microtubules. They also concluded that the nexin linkers are needed to keep the cilium together when it is trying to bend.

These clues have led to a model for how the cilium works (see Figure 3–2). Imagine several smokestacks made of tuna cans that are tightly held together. The tuna can smokestacks are connected by slack wires. Attached to one smokestack is a little motor with an arm that reaches out and holds on to a tuna can in a neighboring smokestack. The motor arm pushes the second smokestack down, sliding it past the first one. As the smokestacks slide past each other, the slack wires begin to stretch and become taut. As the motor arm pushes more, the strain from the wire makes the smokestacks bend. Thus the sliding motion has been converted into a bending motion. Now, let's translate the analogy into biochemical terms. The dynein arms on one microtubule attach to a second, neighboring microtubule, and the dynein uses the biological energy of ATP to "walk up" its neighbor. When this happens the two microtubules begin to slide past each other. In the absence of nexin, they would continue to slide until they separated; however, the protein cross-links prevent neighboring microtubules from sliding by more than a short distance. When the flexible nexin linkers have been elongated to their limit, further walking by dynein makes the nexin linkers tug on the microtubules. As dynein continues its walk, strain increases. Fortunately the microtubules are somewhat flexible, so the dynein-induced sliding motion is converted to a bending motion. ❑

Now, let us sit back, review the workings of the cilium, and consider what they imply. What components are needed for a cilium to work? Ciliary motion certainly requires microtubules; otherwise, there would be no strands to slide. Additionally it requires a motor, or else the microtubules of the cilium would lie stiff and motionless. Furthermore, it requires linkers to tug on neighboring strands, converting the sliding motion into a bending motion, and preventing the structure from falling apart. All of these parts are required to perform one function: ciliary motion. Just as a mousetrap does not work unless all of its constituent parts are present, ciliary motion simply does not exist in the absence of microtubules, connectors, and motors. Therefore we can conclude that the cilium is irreducibly complex—an enormous monkey wrench thrown into its presumed gradual, Darwinian evolution.

The fact that the cilium is irreducibly complex should surprise no one. Earlier in this chapter we saw that a swimming system requires a paddle to contact the water, a motor or source of energy, and a connector to link the two. All systems that move by paddling—ranging from my daughter's toy fish to the propeller of a ship—fail if any one of the components is absent. The cilium is a member of this class of swimming systems. The microtubules are the paddles, whose surface contacts the water and pushes against it. The dynein arms are the motors, supplying the force to move the system. The nexin arms are the connectors, transmitting the force of the motor from one microtubule to its neighbor.[2]

The complexity of the cilium and other swimming systems is *inherent in the task itself.* It does not depend on how large or small the system is, whether it has to move a cell or move a ship: in order to paddle, several components are required. The question is, how did the cilium arise?

AN INDIRECT ROUTE

Some evolutionary biologists—like Richard Dawkins—have fertile imaginations. Given a starting point, they almost always can spin a story to get to any biological structure you wish. The talent can be valuable, but it is a two-edged sword. Although they might think of possible evolutionary routes other people overlook, they also tend to ignore details and roadblocks that would trip up their scenarios. Science, however, cannot ultimately ignore relevant details, and at the molecular level all the "details" become critical. If a molecular nut or bolt is missing, then the whole system can crash. Because the cilium is irreducibly complex, no direct, gradual route leads to its production. So an evolutionary story for the cilium must envision a circuitous route, perhaps adapting parts that were originally used for other purposes. Let's try, then, to imagine a plausible indirect route to a cilium using pre-existing parts of the cell.

To begin, microtubules occur in many cells and are usually used as mere structural supports, like girders, to prop up cell shape. Furthermore, motor proteins also are involved in other cell functions, such as transporting cargo from one end of the cell to another. The motor proteins are known to travel along microtubules, using them as little highways to get from one point to another. An indirect evolutionary argument might suggest that at some point several microtubules stuck together, maybe to reinforce some particular cell shape. After that, a motor protein that normally traveled on microtubules might have accidentally acquired the ability to push two neighboring microtubules, causing a slight bending motion that somehow helped the organism survive. Further small improvements gradually produced the cilium we find in modern cells.

Intriguing as this scenario may sound, though, critical details are overlooked. The question we must ask of this indirect scenario is one for which many evolutionary biologists have little patience: but how *exactly*?

For example, suppose you wanted to make a mousetrap. In your garage you might have a piece of wood from an old Popsicle stick (for the platform), a spring from an old wind-up clock, a piece of metal (for the hammer) in the form of a crowbar, a darning needle for the holding bar, and a bottle cap that you fancy to use as a catch. But these pieces couldn't form a functioning mousetrap without extensive modification, and while the modification was going on, they would be unable to work as a mousetrap. Their previous functions make them ill-suited for virtually any new role as part of a complex system.

In the case of the cilium, there are analogous problems. The mutated protein that accidentally stuck to microtubules would block their function as "highways" for transport. A protein that indiscriminately bound microtubules together would disrupt the cell's shape—just as a building's shape would be disrupted by an erroneously placed cable that accidentally pulled together girders supporting the building. A linker that strengthened microtubule bundles for structural supports would tend to make them inflexible, unlike the flexible linker nexin. An unregulated motor protein, freshly binding to microtubules, would push apart microtubules that should be close together. The incipient cilium would not be at the cell surface. If it were not at the cell surface, then internal beating could disrupt the cell; but even if it were at the cell surface, the number of motor proteins would probably not be enough to move the cilium. And even if the cilium moved, an awkward stroke would not necessarily move the cell. And if the cell did move, it would be an unregulated motion using energy and not corresponding to any need of the cell. A hundred other difficulties would have to be overcome before an incipient cilium would be an improvement for the cell.

SOMEBODY MUST KNOW

The cilium is a fascinating structure that has intrigued scientists from many disciplines. The regulation of its size and structure interests biochemists; the dynamics of its power stroke fascinate biophysicists; the expression of the many separate genes coding for its components engrosses the minds of molecular biologists. Even physicians study them, because cilia are medically important: they occur in some infectious microorganisms, and cilia in the lungs get clogged in the genetic disease cystic fibrosis. A quick electronic search of the professional literature shows more than a thousand papers in the past several years that have *cilia* or a similar word in the title. Papers have appeared on related topics in almost all the major biochemistry journals, including *Science, Nature, Proceedings of the National Academy of Sciences, Biochemistry, Journal of Biological Chemistry, Journal of Molecular Biology, Cell,* and numerous others. In the past several decades, probably ten thousand papers have been published concerning cilia.

Since there is such a large literature on the cilium, since it is of interest to such diverse fields, and since it is widely stated that the theory of evolution is the basis of all modern biology, then one would expect that the evolution of the cilium would be the subject of a significant number of papers in the professional literature. One might also expect that, although perhaps some details would be harder to explain than others, on the whole science should have a good grasp of how the cilium evolved. The intermediate stages it probably went through, the problems that it would encounter at early stages, the possible routes around such problems, the efficiency of a putative incipient cilium as a swimming system—all of these would certainly have been thoroughly worked over. In the past two decades, however, only two articles even attempted to suggest a model for the evolution of the cilium that takes into account real mechanical considerations. Worse, the two papers disagree with each other even about the general route such an evolution might take. Neither paper discusses crucial quantitative details, or possible problems that would quickly cause a mechanical device such as a cilium or a mousetrap to be useless.

The first paper, authored by T. Cavalier-Smith, appeared in 1978 in a journal called *BioSystems*.[3] The paper does not try to present a realistic, quantitative model for even one step in the development of a cilium in a cell line originally lacking that structure. Instead it paints a picture of what the author imagines must have been significant events along the way to a cilium. These imaginary steps are described in phrases such as "flagella [long cilia are frequently called "flagella"] are so complex that their evolution must have involved many stages"; "I suggest that flagella initially need not have been motile, but were slender cell extensions"; "organisms would evolve with a great variety of axonemal structures"; and "it is likely that mechanisms of phototaxis [motion toward light] evolved simultaneously with flagella."

The quotations give the flavor of the fuzzy word-pictures typical of evolutionary biology. The lack of quantitative details—a calculation or informed estimation based on a proposed intermediate structure of how much any particular change would have improved the active swimming ability of the organism—makes such a story utterly useless for understanding how a cilium truly might have evolved.

Let me hasten to add that the author (a well-known scientist who has made a number of important contributions to cell biology) didn't intend that the paper should be taken as presenting a realistic model; he was just trying to be provocative. He was hoping to entice other workers with the promise of his model, however vaguely constructed—to goad them into doing some work to flesh out the emaciated skeleton. Such provocation can be an important service in science. Unfortunately, in the intervening years no one has built upon the model.

The second paper, authored nine years later by a Hungarian scientist named Eörs Szathmary and also appearing in *BioSystems*, is similar in many ways to the first paper.[4] Szathmary is an advocate of the idea, championed by Lynn Margulis, that cilia resulted when a type of swimming bacterium called a "spirochete" accidentally attached itself to a eukaryotic cell.[5] The idea faces the considerable difficulty that spirochetes move by a mechanism (described later) that is totally different from that for cilia. The proposal that one evolved into the other is like a proposal that my daughter's toy fish could be changed, step by Darwinian step, into a Mississippi steamboat. Margulis herself is not concerned with mechanical details; she is content to look for general similarities in some components of cilia and bacterial swimming systems. Szathmary attempted to go a little further and actually discuss mechanical difficulties that would have to be overcome in such a scenario. Inevitably, however, his paper (like Cavalier-Smith's) is a simple word-picture that presents an underdeveloped model to the scientific community for further work. It also has failed at provoking such experimental or theoretical work, either by the author or by others.

Margulis and Cavalier-Smith have clashed in print in recent years.[6] Each points out the enormous problems with the other's model, and each is correct. What is fatal, however, is that neither side has filled in any mechanistic details for its model. Without details, discussion is doomed to be unscientific and fruitless. The scientific community at large has ignored both contributions; neither paper has been cited by other scientists more than a handful of times in the years since publication.[7]

The amount of scientific research that has been and is being done on the cilium—and the great increase over the past few decades in our understanding of how the cilium works—lead many people to assume that even if they themselves don't know how the cilium evolved, *somebody* must know. But a search of the professional literature proves them wrong. Nobody knows.

THE BACTERIAL FLAGELLUM

We humans tend to have a rather exalted opinion of ourselves, and that attitude can color our perception of the biological world. In particular, our attitude about what is higher and lower in biology, what is an advanced organism and what is a primitive organism, naturally starts with the presumption that the pinnacle of nature is ourselves. The presumption can be defended by citing human dominance, and also with philosophical arguments. Nonetheless, other organisms, if they could talk, could argue strongly for their own superiority. This includes bacteria, which we often think of as the rudest forms of life.

Some bacteria boast a marvelous swimming device, the flagellum, which has no counterpart in more complex cells.[8] In 1973 it was discovered that some bacteria swim by rotating their flagella. So the bacterial flagellum acts as a rotary propeller—in contrast to the cilium, which acts more like an oar.

❒ The structure of a flagellum (Figure 3–3) is quite different from that of a cilium. The flagellum is a long, hairlike filament embedded in the cell membrane. The external filament consists of a single type of protein, called "flagellin." The flagellin filament is the paddle surface that contacts the liquid during swimming. At the end of the flagellin filament near the surface of the cell, there is a bulge in the thickness of the flagellum. It is here that the filament attaches to the rotor drive. The attachment material is comprised of something called "hook protein." The filament of a bacterial flagellum, unlike a cilium, contains no motor protein; if it is broken off, the filament just floats stiffly in the water. Therefore the motor that rotates the filament-propeller must be located somewhere else. Experiments have demonstrated that it is located at the base of the flagellum, where electron microscopy shows several ring structures occur. The rotary nature of the flagellum has clear, unavoidable consequences, as noted in a popular biochemistry textbook:

> [The bacterial rotary motor] must have the same mechanical elements as other rotary devices: a rotor (the rotating element) and a stator (the stationary element.)[9]

The rotor has been identified as the M ring in Figure 3–3, and the stator as the S ring. ❒

The rotary nature of the bacterial flagellar motor was a startling, unexpected discovery. Unlike other systems that generate mechanical motion (muscles, for example) the bacterial motor does not directly use energy that is stored in a "carrier" molecule such as ATP. Rather, to

FIGURE 3–3

(TOP) DRAWING OF A BACTERIAL FLAGELLUM SHOWING THE FILAMENT,
HOOK, AND THE MOTOR IMBEDDED IN THE INNER AND OUTER CELL
MEMBRANES AND THE CELL WALL. (BOTTOM) ONE PROPOSED MODEL
FOR THE FUNCTIONING OF THE ACID-DRIVEN, ROTARY MOTOR. THE
DRAWING SHOWS THE INTERNAL COMPLEXITY OF THE MOTOR, WHICH IS
NOT DISCUSSED IN THE TEXT.

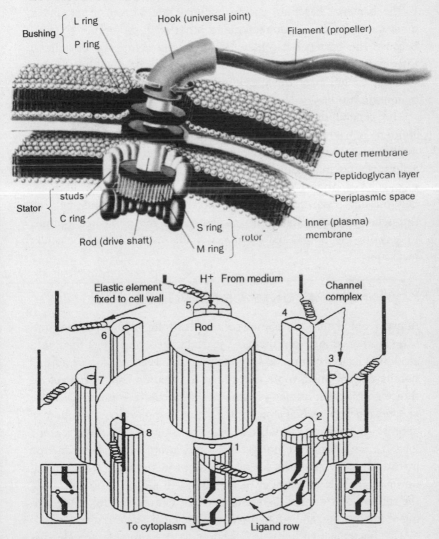

Top from Voet and Voet, fig. 34–84, p. 1259.
Bottom from Caplan, S. R., and Kara-Ivanov, M. (1993), fig. 9A, p. 138. Figures reproduced with permission.

move the flagellum it uses the energy generated by a flow of acid through the bacterial membrane. The requirements for a motor based on such a principle are quite complex and are the focus of active research. A number of models for the motor have been suggested; none of them are simple. (One such model is shown in Figure 3–3 just to give the reader a taste of the motor's expected complexity.)

The bacterial flagellum uses a paddling mechanism. Therefore it must meet the same requirements as other such swimming systems. Because the bacterial flagellum is necessarily composed of at least three parts—a paddle, a rotor, and a motor—it is irreducibly complex. Gradual evolution of the flagellum, like the cilium, therefore faces mammoth hurdles.

The general professional literature on the bacterial flagellum is about as rich as the literature on the cilium, with thousands of papers published on the subject over the years. That isn't surprising; the flagellum is a fascinating biophysical system, and flagellated bacteria are medically important. Yet here again, the evolutionary literature is totally missing. Even though we are told that all biology must be seen through the lens of evolution, no scientist has *ever* published a model to account for the gradual evolution of this extraordinary molecular machine.

IT ONLY GETS WORSE

Above I noted that the cilium contains tubulin, dynein, nexin, and several other connector proteins. If you take these and inject them into a cell that lacks a cilium, however, they do not assemble to give a functioning cilium. Much more is required to obtain a cilium in a cell. A thorough biochemical analysis shows that a cilium contains over *two hundred different kinds of proteins;* the actual complexity of the cilium is enormously greater than what we have considered. All of the reasons for such complexity are not yet clear and await further experimental investigation. Other tasks for which the proteins might be required, however, include attachment of the cilium to a base structure inside the cell; modification of the elasticity of the cilium; control of the timing of the beating; and strengthening of the ciliary membrane.

The bacterial flagellum, in addition to the proteins already discussed, requires about forty other proteins for function. Again, the exact roles of most of the proteins are not known, but they include signals to turn the motor on and off; "bushing" proteins to allow the flagellum to penetrate through the cell membrane and cell wall; proteins to assist in the assembly of the structure; and proteins to regulate the production of the proteins that make up the flagellum.

In summary, as biochemists have begun to examine apparently simple structures like cilia and flagella, they have discovered staggering complexity, with dozens or even hundreds of precisely tailored parts. It is very likely that many of the parts we have not considered here are required for any cilium to function in a cell. As the number of required parts increases, the difficulty of gradually putting the system together skyrockets, and the likelihood of indirect scenarios plummets. Darwin looks more and more forlorn. New research on the roles of the auxiliary proteins cannot simplify the irreducibly complex system. The intransigence of the problem cannot be alleviated; it will only get worse. Darwinian theory has given no explanation for the cilium or flagellum. The overwhelming complexity of the swimming systems push us to think it may never give an explanation.

As the number of systems that are resistant to gradualist explanation mounts, the need for a new kind of explanation grows more apparent. Cilia and flagella are far from the only problems for Darwinism. In the next chapter I will look at the biochemical complexity underlying the apparent simplicity of blood clotting.

CHAPTER 4

RUBE GOLDBERG IN THE BLOOD

SATURDAY MORNING CARTOONS

The name of Rube Goldberg—the great cartoonist who entertained America with his silly machines (Figure 4–1)—lives on in our culture, although the man himself has pretty much faded from view. I was introduced to the notion of a Rube Goldberg machine as a kid watching Saturday morning cartoons. My favorite cartoon was the Bugs Bunny show, and I always enjoyed the loud-mouthed rooster Foghorn Leghorn. I remember a number of episodes in which Foghorn Leghorn would be stuck baby-sitting some smart young chicken with thick glasses while his widowed mother (usually rich) went shopping. At some point Foghorn would annoy the youngster, who would then plot his revenge. A brief scene would show the perturbed chick scribbling some equations on a piece of paper. This got across just how smart he was (after all, you have to be pretty smart to scribble equations) and was an omen that the revenge would be exacted in a precise, scientific way.

A scene or two later Foghorn would be walking along, notice a dollar bill or some other bait on the ground, and pick it up. The dollar was tied by a string to a stick that was propped against a ball. When

FIGURE 4–1   A RUBE GOLDBERG MACHINE.

Mosquito Bite Scratcher

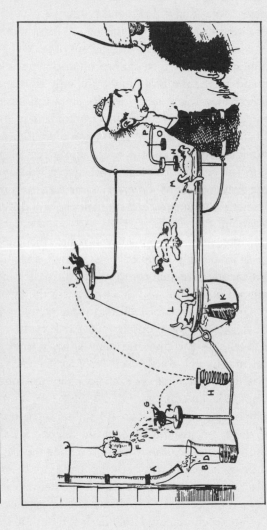

Water from drain-pipe (**A**) drops into flask (**B**)–cork (**C**) rises with water carrying needle (**D**) with it–needle punctures paper tumbler (**E**) containing beer (**F**)–beer sprinkles over bluebird (**G**) and he becomes intoxicated and falls on spring (**H**), which bounces him to platform (**I**)–he pulls string (**J**) thinking it is a worm–string fires off cannon (**K**) which frightens peace-hound (**L**), causing him to jump in air, landing on back in position (**M**)–his heavy breathing raises disc (**N**), which is brought back into its original position by weight (**O**),–the continual breathing of the dog moves scratcher (**P**) up and down over mosquito bite, causing no embarrassment while talking to a lady.

the dollar bill was moved, the attached string pulled down the stick, and the ball would start to roll away as Foghorn stared slack-jawed at the developing action. The ball then would fall off a cliff onto the raised end of a seesaw, smacking it down and sending a rock with an attached piece of sandpaper hurtling into the air. On its upward journey the sandpaper would strike a match sticking out of the cliff, which lit the fuse to a cannon. The cannon would fire; on its downward track the cannonball would hit the rim of a funnel (the only allowance for error in the whole scenario), roll around the edge a few times, and fall through. As it came out of the funnel, the cannonball would hit against a lever that started a circular saw. The saw would cut through a rope, which was holding up a telephone pole. Slowly the telephone pole would begin to fall, and too late Foghorn Leghorn would realize that the fascinating show was at his expense. As he turns to run, the very tip of the telephone pole smacks him on the head and drives him like a peg into the ground.

When you think about it for a moment, you realize that the Rube Goldberg machine is irreducibly complex. It is a single system composed of several interacting parts that contribute to the basic function, and where the removal of any one of the parts causes the system to cease functioning. Unlike the examples of irreducible complexity discussed in previous chapters—the mousetrap, the eukaryotic cilium, and the bacterial flagellum—the cartoon system is not a single piece where the components simultaneously exert force against each other. Rather, it is composed of separate pieces each acting in turn, one after the other, to accomplish its function.

Because the components of the cartoon system are separated from each other in time and space, just one of them (the telephone pole) accomplishes the ultimate purpose of the system (bopping the victim on the head). Nonetheless, the complexity of the system is not thereby reduced, because all system components are required to deliver the blow at the correct time and the correct place. If the mechanism to trigger its fall were not in place, Foghorn could walk back and forth in front of the telephone pole all day and no harm would come to him.

Just as one can catch a mouse with a glue trap instead of a mechanical trap, there are other systems that can deliver a crushing blow to Foghorn Leghorn. You could use a baseball bat, or chop the pole down with an ax while Foghorn was standing in the right place. You could use a nuclear bomb instead of a pole, or attach the string on the bait directly to a shotgun. But none of these other systems are Darwinian precursors to the system used in the cartoon. For example, suppose the string were attached to a dollar bill and directly to the cannon, which would then blast the rooster when he picked up the bait. A Darwinian transformation of that simpler system into the more complex system in the cartoon would require gradually repositioning the cannon, pointing it in a different direction, removing the string from the cannon, reattaching it to the stick, and adding the other paraphernalia. Clearly, however, the system therefore would be out of commission much of the time, so a step-by-step Darwinian transformation is not possible.

Rube Goldberg systems always get a good laugh; the audience enjoys watching the contraption work and appreciates the humor in applying great gobs of ingenuity to a silly purpose. But sometimes a complicated system is used for a serious purpose. In this case the humor fades, but admiration for the delicate interactions of the components remains.

Modern biochemists have discovered a number of Rube Goldberg–like systems as they probe the workings of life on the molecular scale. In the biochemical systems the string, stick, ball, seesaw, rock, sandpaper, match, fuse, cannon, cannonball, funnel, saw, rope, and telephone pole of the cartoon are replaced by proteins with eye-glazing names such as "plasma thromboplastin antecedent" or "high-molecular-weight kininogen." The inner balance and crisp functioning, however, are the same.

OF MILK CARTONS AND CUT FINGERS

When Charles Darwin was climbing the rocks of the Galapagos Islands—pursuing the finches that would eventually bear his name—he must have cut his finger occasionally or scraped a knee. Young adventurer that he was, he probably paid no attention to the little stream of blood trickling out. Pain was a fact of life to the intrepid island explorer, and it had to be borne patiently if any work were to get done.

Eventually the blood would have stopped flowing, and the cut would have healed. If Darwin noticed, it would not have done him much good to speculate about what was going on. He didn't have enough information to even guess at the underlying mechanism of clot formation; the discovery of the structure of the molecules of life lay more then a century in the future. Darwin was an intellectual giant and a great innovator, but no one can guess the future, especially in its critical details.

Blood behaves in a peculiar way. When a container of liquid—like a carton of milk, or a tank truck filled with gasoline—springs a leak, the fluid drains out. The rate of flow can depend on the thickness of the liquid (for example, maple syrup will leak more slowly than alcohol), but eventually it all comes out. No active process resists it. In contrast, when a person suffers a cut it ordinarily bleeds for only a short time before a clot stops the flow; the clot eventually hardens, and the cut heals over. Blood clot formation seems so familiar to us that most people don't give it much thought. Biochemical investigation, however, has shown that blood clotting is a very complex, intricately woven system consisting of a score of interdependent protein parts. The absence of, or significant defects in, any one of a number of the components causes the system to fail: blood does not clot at the proper time or at the proper place.

Some tasks leave little room for error. For example, the most frightening part of an airplane ride for me is the landing. Much of the fear comes from knowing that the plane has to skip over the houses or trees that often are near an airport, and also from realizing that the plane has to stop before it goes off the end of the runway. A few years ago a plane skidded off a runway at LaGuardia Airport into Long Island Sound, killing several people; and it seems that headlines frequently tell of planes crashing just short of the runway. If runways were twenty miles long instead of one mile, I for one would feel more secure.

The landing of an airplane is just one example of a system that has to work within very tight restrictions to avoid disaster. Even the Wright brothers had to worry about landing properly. A little too short or a little too long on the landing, or aiming a little too low or a little too high, and the plane and passengers are in big trouble. But imagine the greater difficulty of landing a plane on autopilot—with no conscious agent to guide it! Blood clotting is on autopilot, and blood clotting requires extreme precision. When a pressurized blood circulation system is punctured, a clot must form quickly or the animal will bleed to death. If blood congeals at the wrong time or place, though, then the clot may block circulation as it does in heart attacks and strokes. Furthermore, a clot has to stop bleeding all along the length of the cut, sealing it completely. Yet blood clotting must be confined to the cut or the entire blood system of the animal might solidify, killing it. Consequently, the clotting of blood must be tightly controlled so that the clot forms only when and where it is required.

PATCHWORK

Over the next few pages you will meet the score of protein players in the game of blood clotting and learn a bit about their roles. Like members of a sports team, some of the players have strange names. Don't worry if the names or the roles of the protein quickly slip your mind— the purpose of the discussion is not for you to memorize trivia. (Besides, the names and relationships will all be shown in Figure 4–3.) Rather, my purpose is to help you get a feel for the complexity of blood clotting and to determine if it could have arisen step by step.

❐ About 2 to 3 percent of the protein in blood plasma (the part that's left after the red blood cells are removed) consists of a protein complex called fibrinogen.[1] The name *fibrinogen* is easy to remember because the protein makes "fibers" that form the clot. Yet fibrinogen is only the potential clot material. Like the telephone pole before it is felled in the story about Foghorn Leghorn, fibrinogen is a weapon waiting to be unleashed. Almost all of the other proteins involved in blood clotting control the timing and placement of the clot. This too is similar to our cartoon example: all components except the telephone pole were required to control the pole's fall.

Fibrinogen is a composite of six protein chains, containing twin pairs of three different proteins. Electron microscopy has shown that fibrinogen is a rod-shaped molecule, with two round bumps on each end of the rod and a single round bump in the middle. So fibrinogen resembles a set of barbells with an extra set of weights in the middle of the bar.

Normally fibrinogen is dissolved in plasma, like salt is dissolved in ocean water. It floats around, peacefully minding its own business, until a cut or injury causes bleeding. Then another protein, called thrombin, slices off several small pieces from two of the three pairs of protein chains in fibrinogen. The trimmed protein—now called fibrin[2]—has sticky patches exposed on its surface that had been covered by the pieces that were cut off. The sticky patches are precisely complementary to portions of other fibrin molecules. The complementary shapes allow large numbers of fibrins to aggregate with each other, like the tubulin–tuna cans from Chapter 3. Just as tubulin does not aggregate to form a random glob but forms a smokestack, however, neither do fibrins stick randomly. Because of the shape of the fibrin molecule, long threads form, cross over each other, and (much as a fisherman's net traps fish) make a pretty protein meshwork that entraps blood cells. This is the initial clot (Figure 4–2). The meshwork covers a large area with a minimum of protein; if it simply formed a lump, much more protein would be required to clog up an area.

FIGURE 4–2

A BLOOD CELL CAUGHT IN THE FIBRIN PROTEIN MESHWORK OF A CLOT.

Manfred Kage/Peter Arnold Inc.

Thrombin, which cuts off the pieces from fibrinogen, is like the circular saw from the Foghorn Leghorn cartoon. Like the saw, thrombin sets in motion the final step of a controlled process. But what if the circular saw ran continuously, without needing the other steps to turn it on? In that case the saw would immediately cut the rope holding up the telephone pole, well before Foghorn moseyed into the vicinity. Similarly, if the only proteins involved in blood coagulation were thrombin and fibrinogen, the process would be uncontrolled. Thrombin would quickly clip all of the fibrinogen to make fibrin; a massive clot would form throughout the animal's circulatory system, solidifying it. Unlike cartoon characters, real animals would rapidly perish. To avoid such an unhappy ending an organism must control the activity of thrombin.

THE CASCADE

❒ The body commonly stores enzymes (proteins that catalyze a chemical reaction, like the cleavage of fibrinogen) in an inactive form for later use. The inactive forms are called proenzymes. When a signal is received that a certain enzyme is needed, the corresponding proenzyme is activated to give the mature enzyme. As with the conversion of fibrinogen to fibrin, proenzymes are often activated by cutting off a piece of the proenzyme that is blocking a critical area. The strategy is commonly used with digestive enzymes. Large quantities can be stored as inactive proenzymes, then quickly activated when the next good meal comes along.

Thrombin initially exists as the inactive form, prothrombin. Because it is inactive, prothrombin can't cleave fibrinogen, and the animal is saved from death by massive, inappropriate clotting. Still, the dilemma of control remains. If the cartoon saw were inactivated, the telephone pole would not fall at the wrong time. If nothing switches on the saw, however, then it would never cut the rope; the pole wouldn't fall even at the right time. If fibrinogen and prothrombin were the only proteins in the blood-clotting pathway, again our animal would be in bad shape. When the animal was cut, prothrombin would just float helplessly by the fibrinogen as the animal bled to death. Because prothrombin cannot cleave fibrinogen to fibrin, something is needed to activate prothrombin. Perhaps the reader can see why the blood-clotting system is called a *cascade*—a system where one component activates another component, which activates a third component, and so on. Since things are beginning to get complicated, it will help a lot to keep track of the discussion with Figure 4–3.

A protein called Stuart factor cleaves prothrombin, turning it into active thrombin that can then cleave fibrinogen to fibrin to form the blood clot.[3] Unfortunately, as you may have guessed, if Stuart factor, prothrombin, and fibrinogen were the only blood-clotting proteins, then Stuart factor would rapidly trigger the cascade, congealing all

FIGURE 4–3

THE BLOOD COAGULATION CASCADE. PROTEINS WHOSE NAMES ARE SHOWN IN NORMAL TYPE FACE ARE INVOLVED IN PROMOTING CLOT FORMATION; PROTEINS WHOSE NAMES ARE ITALICIZED ARE INVOLVED IN THE PREVENTION, LOCALIZATION, OR REMOVAL OF BLOOD CLOTS. ARROWS ENDING IN A BAR INDICATE PROTEINS ACTING TO PREVENT, LOCALIZE, OR REMOVE BLOOD CLOTS.

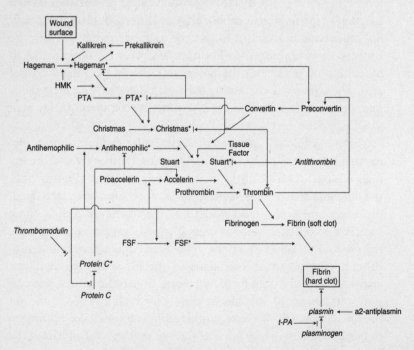

the blood of the organism. So Stuart factor also exists in an inactive form that must first be activated.

At this point there's a little twist to our developing chicken-and-egg scenario. Even activated Stuart factor can't turn on prothrombin. Stuart factor and prothrombin can be mixed in a test tube for longer than it would take a large animal to bleed to death without any noticeable production of thrombin. It turns out that another protein, called accelerin, is needed to increase the activity of Stuart factor. The dynamic duo—accelerin and activated Stuart factor—cleave prothrombin fast enough to do the bleeding animal some good. So in this step we need two separate proteins to activate one proenzyme.

Yes, accelerin also initially exists in an inactive form, called proaccelerin (sigh). And what activates it? Thrombin! But thrombin, as we have seen, is further down the regulatory cascade than proaccelerin. So thrombin regulating the production of accelerin is like having the granddaughter regulate production of the grandmother. Nonetheless, due to a very low rate of cleavage of prothrombin by Stuart factor, it seems there is always a trace of thrombin in the bloodstream. Blood clotting is therefore *auto-catalytic,* because proteins in the cascade accelerate the production of more of the same proteins.

We need to back up a little at this point because, as it turns out, prothrombin as it is initially made by the cell can't be transformed into thrombin, even in the presence of activated Stuart factor and accelerin. Prothrombin must first be modified (not shown in Figure 4–2) by having ten specific amino acid residues, called glutamate (Glu) residues, changed to γ-carboxyglutamate (Gla) residues. The modification can be compared to placing a lower jaw onto the upper jaw of a skull. The completed structure can bite and hang on to the bitten object; without the lower jaw, the skull couldn't hang on. In the case of prothrombin, Gla residues "bite" (or bind) calcium, allowing prothrombin to stick to the surfaces of cells. Only the intact, modified calcium-prothrombin complex, bound to a cell membrane, can be cleaved by activated Stuart factor and accelerin to give thrombin.

The modification of prothrombin does not happen by accident. Like virtually all biochemical reactions, it requires catalysis by a specific enzyme. In addition to the enzyme, however, the conversion of Glu to Gla needs another component: vitamin K. Vitamin K is not a protein; rather, it is a small molecule, like the 11-*cis*-retinal (described in Chapter 1) that is necessary for vision. Like a gun that needs bullets, the enzyme that changes Glu to Gla needs vitamin K to work. One type of rat poison is based on the role that vitamin K plays in blood coagulation. The synthetic poison, called "warfarin" (for the Wisconsin Alumni Research Fund, which receives a cut of the profits from its sale), was made to look like vitamin K to the enzyme that uses it. In the presence of warfarin the enzyme is unable to modify prothrombin. When rats eat food poisoned with warfarin, prothrombin is neither modified nor cleaved, and the poisoned animals bleed to death.

But it still seems we haven't made much progress—now we have to go back and ask what activates Stuart factor. It turns out that it can be activated by two different routes, called the *intrinsic* and the *extrinsic* pathways. In the intrinsic pathway, all the proteins required for clotting are contained in the blood plasma; in the extrinsic pathway, some clotting proteins occur on cells. Let's first examine the intrinsic pathway. (Please follow along using Figure 4–3.)

When an animal is cut, a protein called Hageman factor sticks to the surface of cells near the wound. Bound Hageman factor is then cleaved by a protein called HMK to yield activated Hageman factor. Immediately the activated Hageman factor converts another protein, called prekallikrein, to its active form, kallikrein. Kallikrein helps HMK speed up the conversion of more Hageman factor to its active form. Activated Hageman factor and HMK then together transform another protein, called PTA, to its active form. Activated PTA in turn, together with the activated form of another protein (discussed below) called convertin, switch a protein called Christmas factor to its active form. Finally, activated Christmas factor, together with antihemophilic factor (which is itself activated by thrombin in a manner similar to that of proaccelerin) changes Stuart factor to its active form.

Like the intrinsic pathway, the extrinsic pathway is also a cascade. The extrinsic pathway begins when a protein called proconvertin is turned into convertin by activated Hageman factor and thrombin. In the presence of another protein, tissue factor, convertin changes Stuart factor to its active form. Tissue factor, however, only appears on the outside of cells that are usually not in contact with blood. Therefore, only when an injury brings tissue into contact with blood will the extrinsic pathway be initiated. (A cut plays a role similar to that of Foghorn Leghorn picking up the dollar. It is the initiating event—something outside of the cascade mechanism itself.)

The intrinsic and extrinsic pathways cross over at several points. Hageman factor, activated by the intrinsic pathway, can switch on proconvertin of the extrinsic pathway. Convertin can then feed back into the intrinsic pathway to help activated PTA activate Christmas factor. Thrombin itself can trigger both branches of the clotting cascade by activating antihemophilic factor, which is required to help activated Christmas factor in the conversion of Stuart factor to its active form, and also by activating proconvertin. ☐

Slogging through a description of the blood-clotting system makes a fellow yearn for the simplicity of a cartoon Rube Goldberg machine.

SIMILARITIES AND DIFFERENCES

There are some conceptual differences between Foghorn Leghorn's cartoon contraption and the real-life blood clotting system; the differences emphasize the greater complexity of the biochemical system. The most important contrast is that the clotting cascade has to be turned off at some point before the organism completely solidifies (this will be discussed shortly). A second difference is that the control pathway for blood clotting splits in two. Potentially, then, there are two possible ways to trigger clotting. The relative importance of the two pathways in living organisms is still rather murky. Many experiments on blood clotting are hard to do; some of the proteins—especially the ones involved at the early stages of the pathway—are found in only minute amounts in blood. For example, one hundred gallons of blood contain only about 1 one-thousandth of an ounce of antihemophilic factor. Furthermore, because the initial stages of clotting feed back to generate more of the initial activating proteins, it's often quite difficult to sort out just who is activating whom.

There is also an important conceptual similarity between the

Foghorn attack system and the blood-clotting pathway: both are irreducibly complex. Leaving aside the system before the fork in the pathway, where some details are less well known, the blood-clotting system fits the definition of irreducible complexity. That is, it is a single system composed of several interacting parts that contribute to the basic function, and where the removal of any one of the parts causes the system effectively to cease functioning. The function of the blood clotting system is to form a solid barrier at the right time and place that is able to stop blood flow out of an injured vessel. The components of the system (beyond the fork in the pathway) are fibrinogen, prothrombin, Stuart factor, and proaccelerin. Just as none of the parts of the Foghorn system is used for anything except controlling the fall of the telephone pole, so none of the cascade proteins are used for anything except controlling the formation of a blood clot. Yet in the absence of any one of the components, blood does not clot, and the system fails.

There are other ways to stop blood flow from wounds, but those ways are not step-by-step precursors to the clotting cascade. For example, the body can constrict blood vessels near a cut to help stanch blood flow. Also, blood cells called platelets stick to the area around a cut, helping to plug small wounds. But those systems cannot be transformed gradually into the blood-clotting system any more than a glue trap can be transformed into a mechanical mousetrap.

The simplest blood-clotting system imaginable might be just a single protein that randomly aggregated when the organism was cut. We can liken this to a telephone pole that has been sawed completely through, balancing precariously, depending on the slight vibrations of the ground as Foghorn Leghorn walks by to set it off. The wind or other factors, however, might easily topple the pole when the rooster was not around. Furthermore, the pole is not aimed in any particular direction (such as toward the bait) where Foghorn is likely to be. Similarly, the simplistic clotting system would be triggered inappropriately, causing random damage and wasting resources. Neither the simplified cartoon or clotting "systems" would meet the criterion of minimal function. In Rube Goldberg systems, it is not the final activity (telephone pole falling, clot formation) that is the problem—rather, it is the control system.

One could imagine a blood-clotting system that was somewhat simpler than the real one—where, say, Stuart factor, after activation by the rest of the cascade, directly cuts fibrinogen to form fibrin, bypassing thrombin. Leaving aside for the moment issues of control and timing of clot formation, upon reflection we can quickly see that even such a slightly simplified system cannot change gradually into the more complex, intact system. If a new protein were inserted into the thrombin-less system it would either turn the system on immediately—resulting in rapid death—or it would do nothing, and so have no reason to be selected. Because of the nature of a cascade, a new protein would immediately have to be regulated. From the beginning, a new step in the cascade would require both a proenzyme and also an activating enzyme to switch on the proenzyme at the correct time and place. Since each step necessarily requires several parts, not only is the entire blood-clotting system irreducibly complex, but so is each step in the pathway.

I think a ship canal is a good analogy for this aspect of the blood-clotting system. The Panama Canal allows ships to cross the Isthmus from the Pacific Ocean to the Caribbean Sea. Because the land is higher than sea level, water in a lock lifts a ship up to a level where it can travel along for a while. Then another lock lifts the ship to the next level, and locks on the other side lower the ship back down to sea level. At each lock there is a gate that holds back the water as the ship is raised or lowered; there is also a sluice or water pump that drains or fills the lock. From the beginning each lock must have both features— a gate and a sluice—or it does not function. Consequently, each of the locks along the canal is irreducibly complex. Analogously, each of the control points of the blood-clotting cascade needs both an inactive proenzyme and a separate enzyme to activate it.

IT'S NOT OVER YET

❒ Once clotting has begun, what stops it from continuing until all the blood in the animal has solidified? Clotting is confined to the site of injury in several ways. (Please refer to Figure 4–3.) First, a plasma protein called antithrombin binds to the active (but not the inactive) forms of most clotting proteins and inactivates them. Antithrombin is itself relatively inactive, however, unless it binds to a substance called heparin. Heparin occurs inside cells and undamaged blood vessels. A second way in which clots are localized is through the action of protein C. After activation by thrombin, protein C destroys accelerin and activated antihemophilic factor. Finally, a protein called thrombomodulin lines the surfaces of the cells on the inside of blood vessels. Thrombomodulin binds thrombin, making it less able to cut fibrinogen and simultaneously increasing its ability to activate protein C.

When a clot initially forms, it is quite fragile: if the injured area is bumped the clot can easily be disrupted, and bleeding starts again. To prevent this, the body has a method to strengthen a clot once it has formed. Aggregated fibrin is "tied together" by an activated protein called FSF (for "fibrin stabilizing factor"), which forms chemical cross-links between different fibrin molecules. Eventually, however, the blood clot must be removed after wound healing has progressed. A protein called plasmin acts as a scissors specifically to cut up fibrin clots. Fortunately, plasmin does not work on fibrinogen. Plasmin cannot act too quickly, however, or the wound wouldn't have sufficient time to heal completely. It therefore occurs initially in an inactive form called plasminogen. Conversion of plasminogen to plasmin is catalyzed by a protein called t-PA. There are also other proteins that control clot dissolution, including α_2-antiplasmin, which binds to plasmin, preventing it from destoying fibrin clots. ❑

The cartoon machine that conked Foghorn Leghorn depended critically on the precise alignment, timing, and structure of many components. If the string attached to the dollar bill were too long, or the cannon misaligned, then the whole system would fail. In the same way, the clotting cascade depends critically on the timing and speed at which the different reactions occur. An animal could solidify if thrombin activated proconvertin at the wrong time; it could bleed to death if proaccelerin or antihemophilic factor were activated too slowly. An organism would fade into history if thrombin activated protein C much faster than it activated proaccelerin, or if antithrombin inactivated Stuart factor as fast as it was formed. If plasminogen was activated immediately upon clot formation, then it would quickly dissolve the clot, frustrating the pathway.

The formation, limitation, strengthening, and removal of a blood clot is an integrated biological system, and problems with single components can cause the system to fail. The lack of some blood clotting factors, or the production of defective factors, often results in serious health problems or death. The most common form of hemophilia arises from a deficiency of antihemophilic factor, which helps activated Christmas factor in the conversion of Stuart factor to its active form. Lack of Christmas factor is the second most common form of hemophilia. Severe health problems can also result if other proteins of the clotting pathway are defective, although these are less common. Bleeding disorders also accompany deficiencies in FSF, vitamin K, or α_2-antiplasmin, which are not involved directly in clotting. Additionally, lack of protein C causes death in infancy due to the occurrence of numerous, inappropriate clots.

SHUFFLIN' AROUND

Is it possible that this ultra-complex system could have evolved according to Darwinian theory? Several scientists have devoted much effort to wondering how blood coagulation might have evolved. In the next section you will see what the state-of-the-art explanation is for blood clotting in the professional science literature. But first, there are a few details to attend to.

In the early 1960s it was noticed that some proteins had amino acid sequences that were similar to other proteins' sequences. For example, suppose the first ten amino acids in one protein sequence were ANVLEGKIIS, and in a second protein ANLLDGKIVS. Those two sequences are alike at seven positions and different at three positions. In some proteins, sequences can be similar over hundreds of amino acid positions. To explain the similarity of two proteins it was theorized that in the past a gene was somehow duplicated, and over time the two copies of the gene independently accumulated changes (mutations) in their sequences.[4] After a while there would be two proteins whose sequences were similar, but not identical.

The king of Siam once asked his wise men for a proverb that would be appropriate for any occasion. They suggested "This, too, shall pass." Well, in biochemistry an equally appropriate saying for all occasions is "Things are more complicated than they seem." In the middle 1970s it was shown that genes could occur in pieces. That is, the portion of DNA that coded for the left-hand portion of a protein could be separated along the sequence from portions that coded for the middle, and these could be separated from the DNA that coded for the right-hand portion. It was as if you looked up the word *carnival* in the dictionary and found it listed as "hk*ca*saf*j*rnivckj*ea*lksy." One type of gene might be in one piece; another type might be in dozens of pieces.

The observation of split genes led to the hypothesis that perhaps new proteins could be made by shuffling the DNA fragments of genes that code for parts of old proteins—much as cards can be picked from several piles to give a new arrangement. To support the hypothesis, advocates point to similarities in the amino acid sequences and shapes of discrete portions (called domains) of different proteins.

The proteins of the blood coagulation cascade are often used as evidence for shuffling. Some regions of cascade proteins coded by separate gene pieces have similarities in their amino acid sequences with other regions of the same protein—that is, they are self-similar. Also, there are similarities between regions of different proteins of the cascade. For example, proconvertin, Christmas factor, Stuart factor, and prothrombin all have a roughly similar region of their amino acid sequences. Additionally, in all those proteins the sequence is modified by vitamin K. Furthermore, the regions are similar in sequence to other proteins (not involved with blood coagulation at all) that are also modified by vitamin K.

The sequence similarities are there for all to see and cannot be disputed. By itself, however, the hypothesis of gene duplication and shuffling says nothing about how any particular protein or protein system was first produced—whether slowly or suddenly, or whether by natural selection or some other mechanism. Remember, a mousetrap spring might in some way resemble a clock spring, and a crowbar might resemble a mousetrap hammer, but the similarities say nothing about how a mousetrap is produced. In order to claim that a system developed gradually by a Darwinian mechanism a person must show that the function of the system could "have been formed by numerous successive, slight modifications."

THE STATE OF THE ART

Now we're ready to move forward. In this section I'll reproduce an attempt at an evolutionary explanation of blood clotting offered by Russell Doolittle. What he has done is to hypothesize a series of steps in which clotting proteins appear one after another. Yet, as I will show in the next section, the explanation is seriously inadequate because no reasons are given for the appearance of the proteins, no attempt is made to calculate the probability of the proteins' appearance, and no attempt is made to estimate the new proteins' properties.

Russell Doolittle, a professor of biochemistry at the Center for Molecular Genetics, University of California, San Diego, is the most prominent person interested in the evolution of the clotting cascade. From the time of his Harvard Ph.D thesis, "The Comparative Biochemistry of Blood Coagulation" (1961), Professor Doolittle has examined the clotting systems of different, "simpler" organisms in the hope that that would lead to an understanding of how the mammalian system arose. Doolittle recently reviewed the state of current knowledge in an article in the journal *Thrombosis and Haemostasis*.[5] The journal is intended for professional scientists and doctors of medicine who work on aspects of blood clotting. Essentially, the audience for the journal is those people who know more about blood clotting than anyone else on earth.

Doolittle begins his article by asking the big question: "How in the world did this complex and delicately balanced process evolve? . . . The paradox was, if each protein depended on activation by another, how could the system ever have arisen? Of what use would any part of the scheme be without the whole ensemble?"

These questions go to the heart of this book's inquiry. It is worth quoting Doolittle's article at length. (The reader will find it helpful to refer to Figure 4–3.) I have changed some technical terms in the quote to make it more readable for a general audience.

Blood clotting is a delicately balanced phenomenon involving proteases, antiproteases, and protease substrates. Generally speaking, each forward action engenders some backward-inclined response. Various metaphors can be applied to its step-by-step evolution: action–reaction, point and counterpoint, or good news and bad news. My favorite, however, is yin and yang.

In ancient Chinese cosmology, all that comes to be is the result of combining the opposite principles yin and yang. Yang is the masculine principle and embodies activity, height, heat, light and dryness. Yin, the feminine counterpoint, personifies passivity, depth, cold, darkness and wetness. Their marriage yields the true essence of all things. Keeping in mind that it's only a metaphor, consider the following yin and yang scenario for the evolution of vertebrate clotting. I have arbitrarily designated the enzymes or proenzymes as the yang, and the nonenzymes as the yin.

❐ *Yin:* Tissue Factor (TF) appears as the result of the duplication of a gene for [another protein] that binds EGF domains. The new gene product only comes into contact with the blood or hemolymph after tissue damage.

Yang: Prothrombin appears in an ancient guise with EGF domain(s) attached, the result of a . . . protease gene duplication and . . . shuffling. The EGF domain serves as a site for attachment to and activation by the exposed TF.

Yin: A thrombin-receptor is fashioned by virtue of the duplication of a gene for a [protein region that will stick in a cell membrane]. Cleavage by the TF-activated prothrombin effects cell contractility or clumping.

Yin again: Fibrinogen is born, a bastard protein derived from a thrombin-sensitive [elongated] father and a [protein with a compact structure for a] mother.

Yin again: Antithrombin III appears, the product of a duplication of a [protein with a similar overall structure].

Yang: Plasminogen is generated from the vast inventory of . . . proteases already on hand. It comes with . . . domains that can bind to fibrin. Its activation by binding to bacterial proteins . . . reflects a previous role as an antibacterial agent.

Yin: Antiplasmin arises from the duplication and modification of [a protein with a similar overall structure], probably antithrombin.

Yin and Yang: A thrombin-activatable [cross-linking protein] is unleashed.

Yang: Tissue Plasminogen Activator (TPA) springs forth. Variously shuffled domains allow it to bind to several substances, including fibrin.

Marriage: The modification of prothrombin by the acquisition of a "gla"-domain. The ability to bind calcium and bind to specific [negatively-charged] surfaces is conferred.

Yin: The appearance of proaccelerin[6] as the result of duplicating the [gene for a protein with a similar overall structure] and the acquisition of some other [gene pieces].

Yang: Stuart factor appears, a duplic[ate] of the recently gla-anointed prothrombin; its ability to bind to proaccelerin can bring about . . . activation of prothrombin, independent of the . . . activation by TF.

Yang again: Proconvertin is duplicated from Stuart factor, liberating prothrombin for better binding to fibrin. When combined with tissue factor, proconvertin is able to activate Stuart factor by [cutting it].

Yang again: Christmas factor from Stuart factor. For a period, both bind to proaccelerin.

Yin: Antihemophilic factor from proaccelerin. Quickly adapts to interact with Christmas factor.

Yang: Protein C is genetically derived from prothrombin. Inactivates proaccelerin and antihemophilic factor by limited [cutting].

Divorce: Prothrombin engages in an exchange [of gene pieces] that leaves it with [domains] for binding to fibrin in place of its EGF domains, which are no longer needed for interaction with TF. ❑

HOW'S THAT AGAIN?

Now let's take a little time to give Professor Doolittle's scenario a critical look. The first thing to notice is that no causative factors are cited. Thus tissue factor "appears," fibrinogen "is born," antiplasmin "arises," TPA "springs forth," a cross-linking protein "is unleashed," and so forth. What exactly, we might ask, is causing all this springing and unleashing? Doolittle appears to have in mind a step-by-step Darwinian scenario involving the undirected, random duplication and recombination of gene pieces. But consider the enormous amount of luck needed to get the right gene pieces in the right places. Eukaryotic organisms have quite a few gene pieces, and apparently the process that switches them is random. So making a new blood-coagulation protein by shuffling is like picking a dozen sentences randomly from an encyclopedia in the hope of making a coherent paragraph. Professor Doolittle does not go to the trouble of calculating how many incorrect, inactive, useless "variously shuffled domains" would have to be discarded before obtaining a protein with, say, TPA-like activity.

To illustrate the problem, let's do our own quick calculation. Consider that animals with blood-clotting cascades have roughly 10,000 genes, each of which is divided into an average of three pieces. This gives a total of about 30,000 gene pieces. TPA has four different types of domains.[7] By "variously shuffling," the odds of getting those four domains together[8] is 30,000 to the fourth power, which is approximately one-tenth to the eighteenth power.[9] Now, if the Irish Sweepstakes had odds of winning of one-tenth to the eighteenth power, and if a million people played the lottery each year, it would take an average of about a thousand billion years before *anyone* (not just a particular person) won the lottery. A thousand billion years is roughly a hundred times the current estimate of the age of the universe. Doolittle's casual language ("spring forth," etc.) conceals enormous difficulties. The same problem of ultra-slim odds would trouble the appearance of prothrombin ("the result of a . . . protease gene duplication and . . . shuffling"), fibrinogen ("a bastard protein derived from . . ."), plasminogen, proaccelerin, and each of the several proposed rearrangements of prothrombin. Doolittle apparently needs to shuffle and deal himself a number of perfect bridge hands to win the game. Unfortunately, the universe doesn't have time to wait.

The second question to consider is the implicit assumption that a protein made from a duplicated gene would immediately have the new, necessary properties. Thus we are told that "tissue factor appears as the result of the duplication of a gene for [another protein]." But tissue factor would certainly not appear as the result of the duplication—the other protein would. If a factory for making bicycles were duplicated, it would make bicycles, not motorcycles; that's what is meant by the word *duplication*. A gene for a protein might be duplicated by a random mutation, but it does not just "happen" to also have sophisticated new properties. Since a duplicated gene is simply a copy of the old gene, an explanation for the appearance of tissue factor must include the putative route it took to acquire a new function. This problem is discreetly avoided. Doolittle's scheme runs into the same problem in the production of prothrombin, a thrombin receptor, antithrombin, plasminogen, antiplasmin, proaccelerin, Stuart factor, proconvertin, Christmas factor, antihemophilic factor, and protein C— virtually every protein of the system!

The third problem in the blood-coagulation scenario is that it avoids the crucial issues of how much, how fast, when, and where. Nothing is said about the amount of clotting material initially available, the strength of the clot that would be formed by a primitive system, the length of time the clot would take to form once a cut occurred, what fluid pressure the clot would resist, how detrimental the formation of inappropriate clots would be, or a hundred other such questions. The absolute and relative values of these factors and others could make any particular hypothetical system either possible or (much more likely) wildly wrong. For example, if only a small amount of fibrinogen were available it would not cover a wound; if a primitive fibrin formed a random blob instead of a meshwork, it would be unlikely to stop blood flow. If the initial action of antithrombin were too fast, the initial action of thrombin too slow, or the original Stuart factor or Christmas factor or antihemophilic factor bound too loosely or too tightly (or if they bound to the inactive forms of their targets as well as the active forms), then the whole system would crash. At no step—not even one—does Doolittle give a model that includes numbers or quantities; without numbers, there is no science. When a merely verbal picture is painted of the development of such a complex system, there is absolutely no way to know if it would actually work. When such crucial questions are ignored we leave science and enter the world of Calvin and Hobbes.

Yet the objections raised so far are not the most serious. The most serious, and perhaps the most obvious, concerns irreducible complexity. I emphasize that natural selection, the engine of Darwinian evolution, only works if there is something to select—something that is useful *right now*, not in the future. Even if we accept his scenario for purposes of discussion, however, by Doolittle's own account no blood clotting appears until at least the third step. The formation of tissue factor at the first step is unexplained, since it would then be sitting around with nothing to do. In the next step (prothrombin popping up already endowed with the ability to bind tissue factor, which somehow activates it) the poor proto-prothrombin would also be twiddling its thumbs with nothing to do until, at last, a hypothetical thrombin receptor appears at the third step and fibrinogen falls from heaven at step four. Plasminogen appears in one step, but its activator (TPA) doesn't appear until two steps later. Stuart factor is introduced in one step, but whiles away its time doing nothing until its activator (proconvertin) appears in the next step and somehow tissue factor decides that this is the complex it wants to bind. Virtually every step of the suggested pathway faces similar problems.

Simple words like "the activator doesn't appear until two steps later" may not seem impressive until you ponder the implications.

Since two proteins—the proenzyme and its activator—are both required for one step in the pathway, then the odds of getting both the proteins together are roughly the square of the odds of getting one protein. We calculated the odds of getting TPA alone to be one-tenth to the eighteenth power; the odds of getting TPA and its activator together would be about one-tenth to the thirty-sixth power! That is a horrendously large number. Such an event would not be expected to happen even if the universe's ten-billion year life were compressed into a single second and relived every second for ten billion years. But the situation is actually much worse: if a protein appeared in one step[10] with nothing to do, then mutation and natural selection would *tend to eliminate it.* Since it is doing nothing critical, its loss would not be detrimental, and production of the gene and protein would cost energy that other animals aren't spending. So producing the useless protein would, at least to some marginal degree, be detrimental. Darwin's mechanism of natural selection would actually hinder the formation of irreducibly complex systems such as the clotting cascade.

Doolittle's scenario implicitly acknowledges that the clotting cascade is irreducibly complex, but it tries to paper over the dilemma with a hail of metaphorical references to yin and yang. The bottom line is that clusters of proteins have to be inserted *all at once* into the cascade. This can be done only by postulating a "hopeful monster" who luckily gets all of the proteins at once, or by the guidance of an intelligent agent.

Following Professor Doolittle's example, we could propose a route by which the first mousetrap was produced: The hammer appears as the result of the duplication of a crowbar in our garage. The hammer comes into contact with the platform, the result of shuffling several Popsicle sticks. The spring springs forth from a grandfather clock that had been used as a timekeeping device. The holding bar is fashioned from a straw sticking out of a discarded Coke can, and the catch is unleashed from the cap on a bottle of beer. But things just don't happen that way unless someone or something else is guiding the process.

Recall that Doolittle's audience for the article in *Thrombosis and Haemostasis* are the leaders in clotting research—they know the state of the art. Yet the article does not explain to them how clotting might have originated and subsequently evolved; instead, it just tells a story. The fact is, *no one on earth has the vaguest idea how the coagulation cascade came to be.*

APPLAUSE, APPLAUSE

The preceding discussion was not meant to disparage Russell Doolittle, who has done a lot of fine work over the years in the field of protein structure. In fact he deserves a lot of credit for being one of the very few—possibly the only person—who is actually trying to explain how this complex biochemical system arose. No one else has given this much effort to pondering the origins of blood clotting. The discussion is meant simply to illustrate the enormous difficulty (indeed, the apparent impossibility) of a problem that has resisted the determined efforts of a top-notch scientist for four decades. Blood coagulation is a paradigm of the staggering complexity that underlies even apparently simple bodily processes. Faced with such complexity beneath even simple phenomena, Darwinian theory falls silent.

Like some ultimate Rube Goldberg machine, the clotting cascade is a breathtaking balancing act in which a menagerie of biochemicals—sporting various decorations and rearrangements conferred by modifying enzymes—bounce off one another at precise angles in a meticulously ordered sequence until, at the denouement, Foghorn Leghorn pushes off the telephone pole and gets up from the ground, the bleeding from his wounds stopped. The audience rises to its feet in sustained applause.

CHAPTER 5

FROM HERE TO THERE

THE MEASLES

At the clinic the doctor examines a third young patient who has missed school because of fever, aches, and bloodshot eyes. Like the first two, the boy has the measles. Not rubella. Rubeola. Like the first two, the boy was never immunized. Few kids in the crowded, inner-city neighborhood have been immunized. Measles is rare these days. People forget how dangerous it can be. Parents think of it as a simple matter of temporary freckles and bed rest. They're wrong. Measles makes the patient much more susceptible to other infections. Like encephalitis. The doctor learns that the first patient has just died.

Three cases within a week in the same neighborhood means that the disease is spreading. The doctor fears an epidemic is under way. She immediately calls city health officials and tells them the problem. The health commissioner faxes a request to the Centers for Disease Control (CDC) in Atlanta for ten thousand doses of measles vaccine. The plan is to initiate a crash program of vaccinations in the immediate neighborhood so that spread of the disease will be damped. Infected children will be quarantined; after the outbreak is contained, an educational program will be initiated to alert parents to the abiding dangers of childhood viruses. But first things first: the vaccine is needed immediately.

At the CDC the fax is received, and the request approved. A technician goes down into a storage area where there are a number of large refrigerated rooms stocked with vaccines for measles, smallpox, chicken pox, diphtheria, meningitis, and more. The technician checks the labeling on the packages, sees that the cases in the back corner contain measles vaccine, and loads them onto a cart. He pushes the cart out to a loading dock where a refrigerated truck is waiting to take the packages to the airport. At the airport, the truck glides over to the terminal of a commercial package-delivery service. A number of planes are parked at the terminal, but the truck driver finds a sign marking the plane headed for the right city.

The cases of vaccine are loaded onto the plane, which takes off. At the affected city's airport, another refrigerated truck is there to meet the plane. The packages of vaccine are recognized by their labels, separated from the other packages on the plane, and loaded onto the truck. The driver reads the clinic address from a slip of paper attached to the packages and roars off. At the clinic, a phalanx of medical workers unloads the truck and opens the boxes. Soon a stream of children is entering the clinic to be immunized. As each child passes by, a nurse takes a vial of vaccine, tears off the soft metal cap, inserts the needle of a syringe into the vial, extracts the liquid, and injects it into the arm of the grimacing youngster.

The strategy works. A few more children contract the measles, but no more die. The epidemic is contained, and city officials move on to the educational campaign.

UH-OH

The director leans back in his chair and tosses the script on the table. "Epidemic!"—his first made-for-TV movie—is shaping up pretty well. It has drama, action, cute kids, attractive doctors and nurses, and noble government officials. A killer disease is defeated by human ingenuity, planning, and technical expertise.

Bah! The director does not like happy endings. A cynic down to his toes, he has run across too many stupid, incompetent people to swallow this. His sister's gall bladder was removed by a skilled surgeon;

unfortunately, she had gone into the hospital for an appendectomy. The zoning commission, chaired by a neighbor's uncle, allowed the neighbor to open a video arcade in his quiet neighborhood. And hooligans from the local school let the air out of his tires. The director does not like doctors, hates politicians, and despises kids.

Besides, the director wants to be a great artist. Great artists are supposed to point out human foibles and the tragedies brought on by human limitations. Isn't that what Shakespeare did? They don't pander to the sensibilities of the unwashed masses. So the director closes his eyes and sets to work imagining some different scenarios.

The epidemic begins, officials huddle, and the call goes out to the CDC. The technician goes down to the refrigerated rooms and grabs the boxes labeled "measles vaccine." Onto the truck, into the plane, off to the city, and finally to the clinic. The children noisily file past the nurses and receive their shots. Days pass; three more children die. A week passes, and two dozen children are dead. Some of the dead children had received the vaccine. Two months later, two hundred children are dead, and thousands are sick. Almost all had received the vaccine. Puzzled officials order an investigation, which shows that the packages were mislabeled; the vaccine is for diphtheria, not measles. Almost all of the children in the city are now sick. Nothing can be done. The disease will run its course.

The director smiles. He'll be sure to cast some of the local hooligans as doomed children.

Perhaps, though, the film needs more suspense as the epidemic takes its course. So when the call goes out to the CDC, perhaps the technician goes down to the storage area and sees that all the labels have fallen off the boxes. The refrigerator fan has blown them all around, hopelessly mixing them up. Sweat trickles down the technician's face; he knows that it will take weeks to analyze the boxes to see which vaccine is the right one. During those weeks the disease will spread, politicians will scream, children will die. He may be fired.

Variations on the theme could easily be done. The truck puts the boxes of vaccine on the wrong plane. The plane unloads its cargo into the wrong receiving truck. The truck is hijacked on its way to the clinic. The truck takes the vaccine to the wrong building. The caps on the vaccine bottles are accidentally made from hard metal, not soft, and can't be removed without breaking the bottle and contaminating the vaccine. In all of these cases, the director notes approvingly, human incompetence is highlighted. Great achievements of science— vaccines to conquer disease, airplanes and automobiles to speed supplies on their way—are frustrated by pure, simple stupidity.

The director slaps his knee. Yes, the movie's theme will be a battle, an epic struggle: Albert Einstein versus the Three Stooges. Einstein doesn't have a prayer.

DELIVERY SERVICE

All the problems that cropped up in the director's scenarios concern delivering a package to its final destination. Although the movie showcased death and disease, the same problems are common to all attempts to get a specific package to a specific destination. Suppose you went to a terminal in Philadelphia to catch a bus for New York. A hundred buses were all lined up neatly in a row, motors running, ready to set out to their destinations. But there were no signs on the buses, and the driver and passengers refused to tell you where the bus was headed. So you hopped on board the closest bus and ended up in Pittsburgh.

The bus system has to contend with the same problem that the CDC had: delivering the correct packages (passengers) to the correct destination. The pony express had the same problem. As a rider swooped down to pick up a sack of mail, somebody had to make sure that the mail in the sack was supposed to go to the place where the horse was headed. And the rider had to recognize his destination when he got there.

All cargo delivery systems face common problems: the cargo must be labeled with the correct delivery address; the transporter must recognize the address and put the cargo in the correct delivery vehicle; the vehicle must recognize when it has arrived at the right destination; and the cargo must be unloaded. If any of these steps is missing, then the whole system fails. As we saw in the made-for-TV movie, if the package is mislabeled or no label is present, it doesn't get taken out of the storeroom. If the package is delivered to the wrong address or the container can't be opened once it arrives, then it may as well have never been sent. The entire system must be in place before it works.

Ernst Haeckel thought that a cell was a "homogeneous globule of protoplasm." He was wrong; scientists have shown that cells are complex structures. In particular, eukaryotic cells (which include the cells of all organisms except bacteria) have many different compartments in which different tasks are performed. Just like a house has a kitchen, laundry room, bedroom, and bathroom, a cell has specialized areas partitioned off for discrete tasks (Figure 5–1). These areas include the nucleus (where the DNA resides), the mitochondria (which produce the cell's energy), the endoplasmic reticulum (which processes proteins), the Golgi apparatus (a way station for proteins being transported elsewhere), the lysosome (the cell's garbage disposal unit), secretory vesicles (which store cargo before it must be sent out of the cell), and the peroxisome (which helps metabolize fats). Each compartment is sealed off from the rest of the cell by its own membrane, just as a room is separated from the rest of the house by its walls and door. The membranes themselves can also be considered separate compartments, because the cell places material into membranes that is not found elsewhere.

Some compartments have several discrete sections. For example, mitochondria are surrounded by two different membranes. So a mito-

FIGURE 5–1

THE PARTS OF AN ANIMAL CELL.

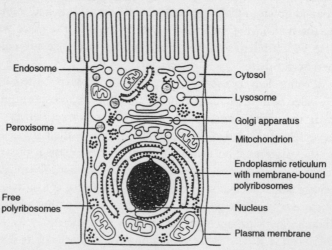

From Alberts et al., fig. 12–1. Reproduced with permission.

chondrion can be thought of as containing four separate sections: the space inside of the inner membrane, the inner membrane itself, the space between the inner and the outer membranes, and the outer membrane itself. Counting membranes and interior spaces, there are more than twenty different sections in a cell.

The cell is a dynamic system; it continually manufactures new structures and gets rid of old material. Since the compartments of a cell are closed off, each area faces the problem of obtaining new materials. There are two ways that it could solve the problem. First, each compartment might make all of its own supplies, like so many self-sufficient villages. Second, new materials could be centrally made and then shipped to other compartments, like a large city making blue jeans and radios to be sent to small towns. Or there might be a mixture of these two possibilities.

In cells, although some compartments make some materials for themselves, the great majority of proteins are centrally made and shipped to other compartments. The shipping of proteins between compartments is a fascinating and intricate process. The details can differ depending on the destination of the protein, just as shipping details can differ depending on whether a package is headed across town or across the ocean. In this chapter I will concentrate on the mechanisms a cell uses to get a protein to the cell's garbage disposal, the lysosome. You will see that the cell must deal with the same problems that the Centers for Disease Control encounters in shipping a vital package.

LOST IN SPACE

A new protein, freshly made in the cell, encounters many molecular machines. Some of the machines grab hold of the protein and send it along to the location it is destined to reach. In a little while I will follow a protein along one pathway from start to finish. Protein machines all have rather exotic names, however, and it is difficult for many people to picture these things in their minds if they are not used to thinking about them. So I will first use an analogy, which will take the next several pages.

The time is far in the future. Humanity has tried to explore space firsthand, but between comets, magnetic storms, and marauding aliens, the dangers were too great. So the job has been given to mechanical space probes that have been shot out into the cosmos to explore the outer edges of our galaxy and beyond. Of course, it takes awhile to get to the edge of the galaxy, and even longer to get beyond, so the space probes have been built to be self-sufficient. They can set down on barren planets and mine for raw materials; they can manufacture brand new machines from ore; and they can capture the energy in starlight and use it to charge their batteries.

The space probe is a machine, so it has to accomplish all of its tasks by painfully detailed mechanisms, not magic. One task is to recycle old batteries; batteries go bad after awhile, so the probe makes new ones. The new batteries are made by grinding up old batteries, recovering the old components, melting them down, recasting the casing, and adding fresh chemicals. One of the machines that is used in this process is called the "battery crusher."

The space probe is shaped like a huge sphere. Inside the sphere are a number of smaller, self-contained spheres, each of which holds machinery for specialized tasks. In the biggest of the interior spheres—let's call it the "library"—are the blueprints for making all the machines in the space probe. These are not ordinary blueprints, however. They can be thought of as blueprints in braille—or perhaps as sheet music for a player piano—where physical indentations in the blueprint cause a master machine to make the machine for which the blueprint codes.

One fine day the space probe senses (by some mechanism we'll ignore) that it needs to make another battery crusher and to send the newly made machine to work in the garbage treatment room, where it will help in recycling old batteries. So the process to do that is set in motion: The blueprint for the battery crusher is photocopied in the library, and the blueprint copy floats over to a window in the library (remember, there's no gravity). On the edge of the blueprint are punch holes arranged in a special pattern, which exactly matches pegs on a scanner mechanism at the window. When the blueprint hooks onto the scanner, the window door opens like the shutter of a camera. The blueprint jiggles loose of the scanner and floats out of the library into the main area of the probe.

In the main area are many machines and machine parts; nuts, bolts, and wires float freely about. In this section reside many copies of what are called master machines, whose job it is to make other machines. They do this by reading the punch holes in a blueprint, grabbing nuts, bolts, and other parts that are floating by, and mechanically assembling the machine piece by piece.

The blueprint for the battery crusher, floating in the main area, quickly comes in contact with a master machine. Whirring, turning appendages on the master machine grab some nuts and bolts and start assembling the crusher. Before it assembles the body of the crusher, however, the master machine first makes a temporary "ornament" that marks the crusher as a machine that has to leave the main area.

In the main area is another machine, called a guide. The shape of the guide is exactly complementary to the shape of the ornament, and little magnets on the guide allow it to attach securely. As the guide snuggles up to the ornament it pushes down on the master machine's switch, causing the master machine to halt its construction of the crusher.

On the outside of one of the interior spheres (we'll call the sphere "processing room #1") is a receiving site that has a shape complementary to part of the guide and part of the ornament. When the guide, ornament, and attached parts bump into that shaped section, the master machine's switch is flipped back on, causing construction of the crusher to resume.

Right next to that shaped section is a window. When the ornament taps on the window (there's a lot of jostling going on), it activates a conveyor belt inside the processing room and the conveyor belt pulls the new battery crusher inside the processing room, leaving the master machine, blueprint, and guide on the outside.

As the crusher was being pulled through the window another machine removed the now-unnecessary ornament. Now, amazingly, constriction machines embedded in the flexible walls of processing room #1 cause a section of the wall to close in on and surround some of the machines, forming a new, free-floating subroom. The remainder of the wall that was left behind smoothly seals itself.

The subroom now floats a short distance through the main area before bumping into a second processing room. The subroom merges with the wall, and spills its contents into processing room #2. The battery crusher then passes through processing rooms #3 and #4 by mechanisms similar to those that took it from room #1 to room #2. It is in the processing rooms that machines receive the tags that direct them to their final destinations. An antenna is placed on the battery crusher and quickly trimmed down to make a very special configuration; the special shape of the trimmed antenna will tell other mechanisms to direct the crusher to the garbage treatment room.

In the wall of the last processing room are machines ("haulers") with a shape complementary to that of the trimmed antenna of the battery crusher. The crusher sticks to the haulers, and that area of the wall begins to pinch off to form a subroom. Outside the subroom is another machine (the "delivery coder") with a shape that exactly complements the shape of a machine (the "port marker") sticking out of the garbage treatment room. The sub-room hooks up to the garbage treatment room through the two complementary machines. Another machine (the "gateway") then drifts by. The gateway has a shape that is complementary to a portion of the delivery coder and the port marker. When it sticks to them the gateway punches a small hole in the garbage treatment room, and the transit sphere merges with it, dumping its contents into the disposal. The battery crusher is finally able to begin its work.

Perhaps by this point in the book, the reader can easily see how the transport system that sent the battery crusher to its destination is irreducibly complex. If any of its numerous components is missing, then the crusher is not delivered to the garbage treatment room. Furthermore, the delicate balance of the system must be maintained; each of the many components that interlock must do so precisely and then disengage, and each must arrive and depart at the proper times. Any single error will cause the system to fail.

REALITY CHECK

This is science fiction, isn't it? Things this complex don't exist in nature, do they? The cell is a "homogeneous globule of protoplasm," isn't it? Well, no, yes, and no.

All of the fantastic machines in our space probe have direct counterparts in the cell. The space probe itself is the cell, the library is the nucleus, the blueprint is the DNA, the copy of the blueprint is RNA, the window of the library is the nuclear pore, the master machines are ribosomes, the main area is the cytoplasm, the ornament is the signal sequence, the battery crusher is a lysosomal hydrolase, the guide is the signal recognition particle (SRP), the receiving site is the SRP receptor, processing room 1 is the endoplasmic reticulum (ER), processing rooms 2 through 4 are the Golgi apparatus, the antenna is a complex carbohydrate, the sub-rooms are coatomer or clathrin-coated vesicles, and various proteins play the roles of the trimmer, hauler, delivery coder, port marker, and gateway. The garbage treatment room is the lysosome.

Let's quickly run through a description of how a protein that is synthesized in the cytoplasm eventually finds its way to the lysosme. This will take just one paragraph. Don't worry if you rapidly forget the names and procedures of cellular transport; the purpose is simply to give you a glimpse of the cell's complexity.

❒ An RNA copy (called messenger RNA, or just mRNA) is made of the DNA gene coding for a protein that works in the cell's garbage disposal—the lysosome. We'll call the protein "garbagease." The mRNA is made in the nucleus, then floats over to a nuclear pore. Proteins in the pore recognize a signal on the mRNA, the pore opens, and the mRNA floats into the cytoplasm. In the cytoplasm the cell's "master machines"—ribosomes—begin making garbagease using the information in the mRNA. The first part of the growing protein chain contains a signal sequence made of amino acids. As soon as the signal sequence forms, a signal recognition particle (SRP) grabs onto the signal and causes the ribosome to pause. The SRP and associated molecules then float over to an SRP receptor in the membrane of the endoplasmic reticulum (ER) and stick there. This simultaneously causes the ribosome to resume synthesis and a protein channel to open in the membrane. As the protein passes through the channel and into the ER, an enzyme clips off the signal sequence. Once in the ER, garbagease has a large, complex carbohydrate placed on it. Coatomer proteins cause a drop of the ER, containing some garbagease plus other proteins, to pinch off, cross over to the Golgi apparatus, and fuse with it. Some of the proteins are returned to the ER if they contain the proper signal. This happens two more times as the protein progresses through the several compartments of the Golgi. Within the Golgi an enzyme recognizes the signal patch on garbagease and places another carbohydrate group on it. A second enzyme trims the freshly attached carbohydrate, leaving behind mannose-6-phosphate (M6P). In the final compartment of the Golgi, clathrin proteins gather in a patch and begin to bud. Within the clathrin vesicle is a receptor protein that binds to M6P. The M6P receptor grabs onto the M6P of garbageance and pulls it on board before the vesicle buds off. On the outside of the vesicle is a v-SNARE protein that specifically recognizes a t-SNARE on the lysosome. Once docked, NSF and SNAP proteins fuse the vesicle to the lysosome. Garbageance has now arrived at its destination and can begin the job for which it was made. ☐

The fictional space probe is so complicated it hasn't been invented yet, even in a crude way. The authentic cellular system is already in place, and every second of every day, this process happens uncounted billions of time in your body. Science is stranger than fiction.

THE DEMANDS OF THE JOB

Garbageance travels a distance of about one ten-thousandth of an inch on its journey from the cytoplasm to the lysosome, yet it requires the services of dozens of different proteins to ensure its safe arrival. In our imaginary TV movie, the vaccine traveled perhaps a thousand miles from the Centers for Disease Control to the big city where it was needed—a trillion times farther than garbageance traveled. But many of the requirements for transporting the vaccine were the same as those for getting the enzyme from the cytoplasm to the lysosome. The demands are imposed by the type of task to be done; they don't depend on the distance traveled, the type of vehicle used, or the materials out of which the signs are made.

A current textbook distinguishes three methods that the cell uses to get proteins into compartments.[1] The first, where a large gate opens or closes to regulate the passage of proteins through the membrane, is known as *gated transport*. This is the mechanism that regulates the flow of material such as newly-made mRNA between the nucleus and the cytoplasm (or in space-probe language, the flow of the blueprint out of the library into the main area). The second method is *transmembrane transport*. This occurs when a single protein is threaded through a protein channel, as when garbagease passed from the cytoplasm into the ER. The third way is *vesicular transport,* where protein cargo is loaded into containers for shipment, as happened for the trip from the Golgi (the final processing room) to the lysosome (the garbage treatment room).

For our purposes the first two methods can be considered to be the same: they both use portals in a membrane that selectively allow proteins through. In the case of gated transport the portal is quite large, and proteins can pass through in their folded form. In the case of transmembrane transport the portal is smaller, and proteins must be threaded through. But in principle there is no roadblock to expanding or contracting the size of a portal, so these are equivalent. Therefore I will call both of these *gated transport.*

What are the bare, essential requirements for gated transport? Imagine a parking garage that is reserved for persons with diplomatic license plates. In place of a human attendant the garage has a scanner that reads a barcode on the license plate, and if the barcode is correct the garage door opens. A car with diplomatic plates drives up, the scanner scans the barcode, the door opens, and the car drives in. It doesn't matter if the car drove ten feet to the garage or ten thousand miles, or whether the vehicle is a truck, jeep, or motorcycle; if the barcode is there, it can pass through. Thus three basic components are required for gated transport at the garage: an identification tag; a scanner; and a gate that is activated by the scanner. If any of these things are missing, then either the vehicle does not get in or the garage is no longer a reserved area.

Because gated transport requires a minimum of three separate components to function, it is irreducibly complex. And for this reason the putative gradual, Darwinian evolution of gated transport in the cell faces massive problems. If proteins contained no signal for transport, they would not be recognized. If there were no receptor to recognize a signal or no channel to pass through, again transport would not take place. And if the channel were open for all proteins, then the enclosed compartment would not be any different from the rest of the cell.

Vesicular transport is even more complicated than gated transport. Suppose now that, instead of the diplomats' cars entering the garage one at a time, all diplomats had to drive their cars into the back of a large tractor-trailer truck, the truck would drive into the special garage, and the cars would drive off the truck and park. Now we need a way for the truck to recognize the proper cars, a way for the garage to recognize the truck, and a way for the cars to get out of the truck inside the garage. Such a scenario requires six separate components: (1) an identification tag on the cars; (2) a truck that can carry the cars; (3) a scanner on the truck; (4) an identification tag on the truck; (5) a scanner on the garage; (6) an activatable garage gate. In the cell's vesicular transport system these components correspond to mannose-6-phosphate, the clathrin vesicle, the M6P receptor in the clathrin vesicle, v-SNARE, t-SNARE, and SNAP/NSF proteins. In the absence of any of these functions, either vesicular transport cannot take place or the integrity of the destination compartment is compromised.

Because vesicular transport requires several more components than gated transport, it cannot develop gradually from gated transport. For example, if we had barcode stickers on the diplomats' cars, placing cars inside a truck (a vesicle to transport them) would hide the stickers, and they would fail to enter the garage. Or suppose instead that the truck had the same label that the cars had, so it could enter the garage. But we would still be missing a mechanism to get the cars on the truck, so the truck would be of no use. If some cars randomly entered the truck then, again, nondiplomats' cars would enter the garage. Returning to the world of the cell, if a vesicle just "happened" to form there would be no mechanism for identifying the proteins that should enter it, and no way to specify its destination. Placing proteins containing address labels into an unlabeled vesicle would make the labels unavailable, and therefore would be detrimental to the organism that had a happily functioning gated transport system. Gated transport and vesicular transport are two separate mechanisms; neither helps in understanding the other.

The brief sketch of the requirements for gated and vesicular transport in this chapter did not take into account many complexities of the systems. But since these only make the system more intricate, they cannot ameliorate the irreducible complexity of targeted transport.

SECOND-HAND ROSE

Irreducibly complex systems like mousetraps, Rube Goldberg machines, and the intracellular transport system cannot evolve in a Darwinian fashion. You can't start with a platform, catch a few mice, add a spring, catch a few more mice, add a hammer, catch a few more mice, and so on: The whole system has to be put together at once or the mice get away. Similarly, you can't start with a signal sequence and have a protein go a little way towards the lysosome, add a signal receptor protein, go a little further, and so forth. It's all or nothing.

Perhaps, though, we're overlooking something. Perhaps one of the parts of a mousetrap was used for some purpose other than trapping mice, and so were the other parts. At some point several parts that were being used for other purposes suddenly came together to produce a functioning trap. And perhaps the components of the intracellular transport system were originally performing other tasks in the cell, then switched to their present role. Could that happen?

An exhaustive consideration of all possible roles for a particular component can't be done. We can, however, consider a few likely roles for some of the components of the transport system. Doing so shows it is extremely implausible that components used for other purposes fortuitously adapted to new roles in a complex system.

Suppose we start with a protein that because it had an oily region, resided in the cell's membrane. Suppose further that it was beneficial for the protein to be there because it toughened the membrane, making it resistant to tears and holes. Could that protein somehow turn into a gated channel? This is like asking if wooden beams in a wall could be transformed, step by Darwinian step, small mutation by small mutation, into a door with a scanner. Suppose wooden beams were brought together, and the area between them was weakened so much that plaster cracked and a hole formed in the wall. Would that be an improvement? The hole in the wall would let insects, mice, snakes, and other things into the room; it would let heat or air-conditioning out. Similarly, a mutation that caused proteins to aggregate in the membrane, leaving a small hole, would let stored foodstuffs, salt, ATP, and other needed materials float away. That is no improvement. A house with a hole in the wall would never sell, and a cell with a hole in it would be at a great disadvantage compared to other cells.

❐ Suppose instead that a protein could bind to the beginnings of new proteins as they were being put together by the ribosome. Suppose that was an improvement because new, unfolded proteins are more vulnerable, so placing a folded protein on them would protect them until they were fully made and folded. Could such a protein develop into, say, the signal recognition particle (SRP)? No. Such a protein would help a new protein fold rapidly, not keep it unfolded—the opposite of what modern SRP does. Folded· proteins, however, can't get through the gated channel where the modern SRP takes them. Further, if a proto-SRP caused the ribosome to halt its synthesizing, as the modern SRP does, but the machinery to turn the ribosome back on was not yet in place, then that would kill a cell (some deadly poisons kill by turning off the cell's ribosomes). So we have a dilemma: in the beginning an uncontrolled inhibitor of protein synthesis would kill the cell, but a temporary halt in protein synthesis is crucial in modern cells. If the ribosome does not pause, the new protein gets so big that it can't fit through a gated channel. So it appears that the modern SRP could not have developed from a protein whose job it was to bind new proteins and protect them from degradation.

Suppose that an enzyme placed a large carbohydrate group (the "bauble") on proteins as they were made. Suppose that helped stabilize the protein somehow, making it last longer in the cell. Could that step eventually become part of the intracellular transport chain? No. The bauble, because it would make the protein larger, would prevent it from passing through any future gate that looked like a modern gate in the ER. The bauble would actually be a hindrance to developing a transport system.

In the same way, other isolated parts of the system would actually be damaging to the cell, not helpful. An enzyme that clipped off the signal sequence (the "ornament") would be detrimental if the signal sequence was playing a positive role in a primitive cell. Trimming of the bauble would be a step backward if the bauble had a job to do. Trapping of proteins like "garbagease" inside a vesicle would be harmful if garbagease originally had to work in the open. ❑

In Chapter 2 I noted that one couldn't take specialized parts of other complex systems (such as the spring from a grandfather clock) and use them directly as specialized parts of a second irreducible system (like a mousetrap) unless the parts were first extensively modified. Analogous parts playing other roles in other systems cannot relieve the irreducible complexity of a new system; the focus simply shifts from "making" the components to "modifying" them. In either case, there is no new function unless an intelligent agent guides the setup. In this chapter we see that construction of a transport system faces the same problem: the system can't be put together piecemeal from either new or secondhand parts.

DEATH AT AN EARLY AGE

In one version of our made-for-TV movie, a wrong label was placed on a carton of vaccine, and children died. Fortunately, it was only make-believe: a story about a story. But in real life, mixed-up or missing labels can cause real deaths.

A crying two-year-old girl stands in front of a height chart, with the aid of an adult's helping hand. She is only two feet tall. Her face and eyes are puffed up, and her legs are bent. She moves stiffly. She is severely retarded. A medical examination shows an enlarged heart, liver, and spleen. A cough and runny nose bespeak another of the many upper respiratory infections she has endured in her young life. The doctor takes a tissue sample from the girl and sends it to a lab for analysis; a lab worker grows cells from the sample in a Petri dish and examines them under a microscope. Each of the cells contains thousands of little, dense grains that aren't present in normal cells. The grains are called "inclusion bodies"; the little girl has I-cell disease.[2] Because the disease is progressive, the skeletal and neural difficulties will increase with time. The girl will die before the age of five.

I-cell disease is caused by a defect in the protein transport pathway. The cells of patients with the disease lack one of the machines in the long chain that takes proteins from the cytoplasm to the lysosome. Because of the defect, enzymes intended for the lysosome never make it there. Instead they are shunted off in the wrong vesicle to the cell membrane and dumped into the extracellular space.

The cell is a dynamic system, and just as it must build new structures, it must continually degrade old ones. Old material is brought to the lysosome for degradation. In children with I-cell disease, the garbage is dumped into the disposal as it should be, but the disposal is broken: neither "garbagease" nor any other degradative enzyme that normally decomposes old structures is present. As a result garbage piles up, and lysosomes get filled. The cell makes new lysosomes to accomodate the increasing waste, but the new compartments eventually fill up with the detritus of cellular life. Over time the entire cell becomes bloated, tissues become enlarged, and the patient dies.

A child can die because of this single defect in one of the many machines needed for taking proteins to the lysosome. A single flaw in the cell's labyrinthine protein-transport pathway is fatal. Unless the entire system were immediately in place, our ancestors would have suffered a similar fate. Attempts at a gradual evolution of the protein transport system are a recipe for extinction.

Because of the medical problems associated with the failure of the transport system, and because the system is so intricate and fascinating, we might expect the evolutionary development of vesicular protein transport to be a busy area of research. How could such a system develop step-by-step? What hurdles would the cell have to overcome as it moved from some other method of dealing with garbage to a coated vesicle specifically targeted to, and equipped for merger with, the lysosome? Once again, if we looked in the literature for an explanation of the evolution of vesicular transport, we would be crushingly disappointed. Nothing is there.

Annual Review of Biochemistry (or *ARB*)is a book series, very popular with biochemists, that reviews the current state of knowledge in selected research areas. In 1992 an article was published in *ARB* concerning "Vesicle-Mediated Protein Sorting."[3] The authors begin their review by stating the obvious: "The transport of proteins between membrane-bounded organelles is an immensely complex process." They proceed in professional fashion to describe the systems and current research in the area. But we can read from one end of the forty-six-page review to the other without encountering an explanation for how such a system might have gradually evolved. The topic is off the radar screen.

Logging on to a computer database of the professional literature in the biomedical sciences allows you to do a quick search for key words in the titles of literally hundreds of thousands of papers. A search to see what titles have both *evolution* and *vesicle* in them comes up completely empty. Slogging through the literature the old-fashioned way turns up a few scattered papers that speculate on how gated transport between compartments of a eukaryotic cell might have developed.[4] But all the papers assume that the transport systems came from preexisting bacterial transport systems that already had all the components that modern cells have. This does us no good. Although the speculations may have something to do with how transport systems could be duplicated, they have nothing to do with how the initial systems got there. At some point this complex machine had to come into existence, and it could not have done so in step-by-step fashion.

Perhaps the best place to get an overview of vesicle transport is from the textbook *Molecular Biology of the Cell* by National Academy of Science President Bruce Alberts, Nobel Prize winner James Watson, and several more coauthors. The textbook spends 100 pages on the elegant details of gated and vesicular transport.[5] In that 100 pages there is a one-and-a-half-page section entitled "The Topological Relationships of Membrane-Bounded Organelles Can Be Interpreted in Terms of Their Evolutionary Origins." In this section the authors point out that if a vesicle pinches off from the cell membrane and into the cell, then its inside is equivalent to the outside of the cell. They then suggest that the nuclear membrane, ER, Golgi, and lysosomes first arose when parts of the cell membrane pinched off. This may or may not be true, but it does not even address the origin of protein transport, either vesicular or gated. Clathrin is not mentioned in this short section, nor are the problems of loading the correct cargo into the correct vesicle and targeting it to the correct compartment. In short, the discussion is irrelevant to the questions we are asking. At the end of our literature search, we know no more than when we started.

SUMMING UP AND LOOKING AHEAD

Vesicular transport is a mind-boggling process, no less complex than the completely automated delivery of vaccine from a storage area to a clinic a thousand miles away. Defects in vesicular transport can have the same deadly consequences as the failure to deliver a needed vaccine to a disease-racked city. An analysis shows that vesicular transport is irreducibly complex, and so its development staunchly resists gradualistic explanations, as Darwinian evolution would have it. A search of the professional biochemical literature and textbooks shows that no one has ever proposed a detailed route by which such a system could have come to be. In the face of the enormous complexity of vesicular transport, Darwinian theory is mute.

In the next chapter I will examine the art of self-defense—but, of course, on a molecular scale. Just as machine guns, battle cruisers, and nuclear bombs are necessarily sophisticated machines in our larger world, we will see that tiny cellular defense mechanisms are quite complex, too. Few things are simple in Darwin's black box.

CHAPTER 6

A DANGEROUS WORLD

ALL SHAPES AND SIZES

Enemies abound. Paranoia has nothing to do with it; we are surrounded by creatures that, for one reason or another, want to do us in.
Since most people don't want to die just yet, they take steps to defend
themselves.

Threats of aggression can come in all shapes and sizes, so defenses
have to be versatile. The largest-scale threat is war between nations.
Rulers of nations always seem to be wanting the resources of neighboring countries, so threatened countries have to defend themselves or
suffer unpleasant consequences. In modern times, countries can have
very sophisticated means of defense indeed. The United States has
stockpiled atomic bombs; if some other country shakes its proverbial
fist at us, we can rattle our bombs at them. If threats escalate to vio-
lence and we don't wish to use atomic bombs for one reason or another, then other machines can be deployed: jets that drop "smart"
bombs, AWACS planes that monitor the air space for many miles,
tanks equipped for night combat, surface-to-air-missiles that shoot
down surface-to-surface missiles, and much more. To the techno-war-
monger, we live in a golden age.

Big threats like war are important, but other types of aggression can kill, too. Terrorist bombings of planes or gas attacks on subways have, unfortunately, become too frequent for comfort. Worse, none of the weapons mentioned above will help much to prevent a subway gas attack. When the nature of the enemy changes dramatically—from a foreign country to a domestic terrorist group—the nature of the defense must also change. Instead of bombs, government officials install metal detectors at airports and place guards with guns at strategic locations.

Terrorism and war threaten us, but they happen infrequently. On a day-to-day basis more people are assaulted by muggers and mayhem in their neighborhood than by exotic groups or foreign countries. The streetwise city dweller will have bars on his window, use an intercom or peephole to see who is at the door, and carry a can of pepper spray when it's time to walk the dog. In lands where such modern conveniences are unknown, stone or wooden walls can be built around the hut to keep out intruders (both two- and four-footed), and a spear is kept by the bed in case the wall is breached.

A stick, rock, barrier, gun, alarm, tank, and atomic bomb can all be used to help fend off attacks. Since the circumstances in which each weapon is useful might vary considerably, there is a lot of overlap. Both a stick and a pistol can deter a mugger; a pistol and a tank can threaten a terrorist group; and both a tank and an atomic bomb can be used against a foreign country. Looked at this way, we can speak about the "evolution" of defensive systems. We can talk about an arms race in which the equipment of competing sides becomes more and more sophisticated. We can tell stories about life being a struggle where people or countries with the best defenses survive. But before we hop in a box and fly off with Calvin and Hobbes, we need to recall the distinction between conceptual precursors and physical precursors. A rock and a gun can both be used for defense, but a rock cannot be turned into a gun by a series of small steps. A can of pepper spray is not a physical precursor of a hand grenade. A jet plane cannot be changed into an atomic bomb one nut and bolt at a time, even though both the plane and the bomb do contain nuts and bolts. In Darwinian evolution, only physical precursors count.

Humans and large animals are not the only threats a person encounters. There are also Lilliputian aggressors against whom bombs or guns or rocks are ineffective. Bacteria, viruses, fungi—they all would love to eat us if they could. Sometimes they do, but most times they don't because our bodies have an array of defensive systems to deal with microscopic attacks. The first line of defense is the skin. Like a stockade fence, the skin works by a relatively low-tech method: it's a barrier that is hard to breach. Burn victims often succumb to massive infections because the skin barrier has been broken and the internal defenses can't cope with the overwhelming numbers of invaders. But although skin is an important part of the body's defense, it is not a physical precursor of the immune system.

To discourage any outsider who manages to climb to the top, sometimes stockade walls have spikes on them. Where I lived in the Bronx, almost all of the cyclone fences were topped with razor wire, which apparently is more effective at lacerating intruders than old-fashioned barbed wire. Spikes and razor wire are not parts of the fence proper, they are little add-ons that increase the effectiveness of the barrier. Still, like the fence itself, razor wire is not a physical precursor to, say, a gun or a landmine.

Skin, too, has add-ons that increase its effectiveness as a barrier. In a biochemistry laboratory you often have to wear gloves to protect yourself from the material you're handling, but sometimes you have to wear gloves to protect the material from you. People who work with RNA wear gloves because human skin excretes an enzyme that chops up RNA. Why? It turns out that many viruses are made from RNA. To such a virus, the enzyme is like razor wire on the skin: any RNA that tries to breach the barrier gets lacerated.

There are other types of spikes on the skin. One of the most interesting is a class of molecules called magainins, discovered by a biologist named Mike Zasloff after he wondered why live laboratory frogs that are cut open and sewed back up in nonsterile conditions rarely get infections. He showed that their skin excretes a substance which can kill bacterial cells; since then, magainins have been discovered in many kinds of animals. But magainins, like the RNA-destroying enzymes, are not precursors to the sophisticated defense systems under the skin of animals.

To find the heavy weaponry, we have to peek under our skins. The internal defense system of vertebrates is dizzyingly complicated. Like the modern U.S. army, it has a variety of different weapons that can overlap in their use. But like the weapons we discussed above, we must not automatically assume the different parts of the immune system are physical precursors of each other. Although the body's defenses are still an active area of research, much is known in detail about particular aspects. In this chapter I will discuss selected parts of the immune system and point out the problems they present for a model of gradual evolution. Those who become intrigued by the cleverness of the systems and want to know more are encouraged to pick up any immunology text for the details.[1]

THE RIGHT STUFF

When a microscopic invader breaches the outer defenses of the body, the immune system swings into action. This happens automatically. The molecular systems of the body, like the Star Wars anti-missile system that the military once planned, are robots designed to run on autopilot. Since the defense is automated, every step has to be accounted for by some mechanism. The first problem that the automated defense system has is how to recognize an invader. Bacterial cells have to be distinguished from blood cells; viruses have to be distinguished from connective tissue. Unlike us, the immune system can't see, so it has to rely initially on something akin to a sense of touch.

❐ Antibodies are the "fingers" of the blind immune system—they allow it to distinguish a foreign invader from the body itself. Antibodies are formed by an aggregation of four chains of amino acids (Figure 6–1): two identical light chains, and two identical heavy chains. The heavy chains are about twice as big as the light chains. In the cell, the four chains make a complex that resembles the letter Y. Because the two heavy chains are the same and the two light chains are the same, the Y is symmetrical: if you took a knife and cut it down the middle you'd get identical halves, with one heavy and one light chain in each half. At the end of each pronged tip of the Y there is a depression (called a binding site). Lining the binding site are portions of both the light chain and the heavy chain. Binding sites come in a large variety of shapes. One antibody might have a binding site with a piece jutting up here, a hole over there, and an oily patch on the edge. A second antibody might have a positive charge on the left, a crevice in the middle, and a bump on the right.

FIGURE 6–1

SCHEMATIC DRAWING OF AN ANTIBODY MOLECULE.

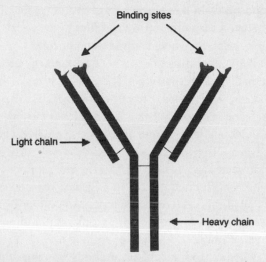

If the shape of a binding site just happens to be exactly complementary to the shape of a molecule on the surface of an invading virus or bacterium, then the antibody will bind to that molecule. To get a feel for it, imagine a household object with a depression in it and a few knobs poking up out of the depression. My youngest daughter has a doll wagon with front and back seats—something like that will do nicely. Now take the wagon/object, go around the house, and see how many other articles will fit snugly into the depression, filling both the front seat and the back seat without leaving any spaces. If you find even one, you're luckier than I am. Nothing in my house fit snugly in the wagon, and neither did anything in my office or laboratory. I imagine there's some object out in the world with a shape complementary to the wagon's, but I haven't found it yet.

The body has a similar problem: the odds of any given antibody binding to any given invader are pretty slim. To make sure that at least one kind of antibody is available for each attacker, we make billions to trillions of them. Usually, for any particular invader, it takes 100,000 to find one antibody that works.

When bacteria invade the body, they multiply. By the time an antibody binds to a bacterium there may be many, many copies of the bug floating around. Against this Trojan horse that breeds, the body has 100,000 guns, but only one works. One handgun isn't going to do much good against a horde; somehow reinforcements have to be brought in. There's a way to do this, but first I have to back up and explain a bit more about where antibodies come from.

There are billions of different kinds of antibodies. Each kind of antibody is made in a separate cell. The cells that make antibodies are called B cells, which is easy to remember because they are produced in the bone marrow.[2] When a B cell is first born, mechanisms inside of it randomly choose one of the many antibody genes that are encoded in its DNA. That gene is said to be "turned on"; all other antibody genes are "turned off." So the cell produces only one kind of antibody, with one kind of binding site. The next cell that's made will in all likelihood have a different antibody gene turned on, so it will make a different protein with a different binding site. The principle, then, is one cell, one type of antibody.

Once a cell commits to making its antibody, you might think that the antibody would leave the cell so it could patrol the body. But if the contents of all B cells were dumped out into the body, there would be no way to tell which cell the antibody came from. The cell is the factory that makes the particular type of antibody; if the antibody finds a bacterium, we need to tell the cell to send us reinforcements. But with this hypothetical setup, we can't get a message back.

Fortunately, the body is smarter than that. When a B cell first makes its antibody, the antibody anchors in the cell membrane with the prongs of the Y sticking out (Figure 6–2). The cell does this trick by using the gene for the normal antibody, and also using a little piece of a gene that codes for an oily tail on the protein. Since the membrane is oily, too, the piece sticks in the membrane. This step is critical, because now the binding site of the antibody is attached to its factory. The entire B cell factory patrols the body; when a foreign invader enters, the antibody-with-attached-cell binds.

Now we have the factory close at hand to the invaders. If the cell could be signaled to make more of the antibody, then the fight would be helped by reinforcements. Fortunately, there is a way to send a signal; unfortunately, it's pretty convoluted. When an anti-

FIGURE 6–2

SCHEMATIC DRAWING OF A B CELL.

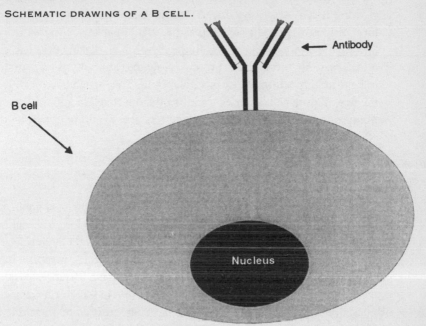

body on a B cell binds to a foreign molecule it triggers a complex mechanism to swallow the invader: in effect, the munitions factory takes a hostage. The antibody then breaks off a piece of membrane to make a little vesicle—a self-made taxicab. In this taxi, the hostage is brought into the B-cell factory. Inside the cell (still in the cab) the foreign protein is chopped up, and a piece of the foreign protein sticks to another protein (called an MHC protein). The cab then returns to the membrane of the cell. Outside the factory, along comes another cell (called a helper T cell). The helper T cell binds to the B cell, which is "presenting" the chopped-up piece of invader (the foreign fragment in the MHC protein) for the T cell's consideration. If the fit is just right, it causes the helper T cell to secrete a substance called interleukin. Interleukin is like a message from the Department of Defense to the munitions factory. By binding to another protein on the surface of the B cell, the interleukin sets off a chain of events that sends a message to the nucleus of the B cell. The message is: grow!

The B cell begins to reproduce at a rapid rate. T cells continue to secrete interleukin if they are bound to a B cell. Eventually the growing B-cell factory produces a series of spinoff factories in the form of specialized cells called 'plasma cells.' Instead of producing a form of the antibody that sticks in the membrane, plasma cells leave off the last oily piece of the protein. Now free antibody is extruded in large amounts into the extracellular fluid. The switch is critical. If the new plasma-cell factories were like the old B-cell factory, the antibodies would all be confined to quarters and would be much less effective at inhibiting the invaders. ❐

STEP BY STEP

Could this system have evolved step-by-step? Consider the vast pool of billions to trillions of factory B cells. The process of picking the right cell out of a mixture of antibody-producing cells is called clonal selection. Clonal selection is an elegant way to mount a specific response in great numbers to a wide variety of possible foreign invaders. The process depends on a large number of steps, some of which I have not discussed yet. Leaving those aside for now, let's ask what the minimum requirements are for a clonal selection system, and if those minimum requirements could be produced step-by-step.

The key to the system is the physical connection of the binding ability of the protein with the genetic information for the protein. Theoretically this could be accomplished by making an antibody where the tail of the Y bound to the DNA that coded for the protein. In real life, however, such a setup wouldn't work. The protein might be connected to its genetic information, but because the cell is surrounded by a membrane, the antibody would never come in contact with the foreign material, which is floating around outside the cell. A system where both the antibody and its attached gene were exported from the cell would overcome that problem, only to run into a different one: outside the cell there would be no cellular machinery to translate the DNA message into more protein.

Anchoring the antibody in the membrane is a good solution to the problem; now the antibody can mix it up with a foreign cell and still be near its DNA. But although the antibody can bind the foreign material without floating away from the cell, it does not have direct physical contact with the DNA. Since the protein and DNA are blind, there must be a way to get a message from one to the other.

Just for now, for the sake of argument, let's forget about the tortuous way that the message of binding actually gets to the B-cell nucleus (requiring the taxicab, ingestion, MHC, helper T cells, interleukin, and so on). Instead let's imagine a simpler system where there's only one other protein. Let's say that when the antibody binds to a foreign molecule, something happens that attracts some other protein—a messenger to take word of a hostage to the factory nucleus. Maybe when the hostage is first found, the shape of the antibody changes, perhaps pulling up a little on the antibody's tail. Perhaps part of the antibody's tail sticks into the inside of the cell, which is what triggers the messenger protein. The change in the tail could cause the messenger protein to scuttle into the nucleus and bind to the DNA at a particular point. Binding to the right place on the DNA is what causes the cell to start growing and to start producing antibody without the oily tail—antibody that gets sent out of the cell to fight the invasion.

Even in such a simplified scheme, we are left with three critical ingredients: (1) the membrane-bound form of the antibody; (2) the messenger; and (3) the exported form of the antibody. If any of these components is missing, the system fails to function. If there is no antibody in the membrane, then there's no way to connect a successful antibody that binds a foreign invader to the cell containing the genetic information. If there is no exported form of the antibody, then when the signal is received there is nothing to send out into the world to fight. If there is no messenger protein, then there is no connection between binding the membrane antibody and turning on the right gene (making the system about as useful as a doorbell whose wires had been cut).

A cell hopefully trying to evolve such a system in gradual Darwinian steps would be in a quandary. What should it do first? Secreting a little bit of antibody into the great outdoors is a waste of resources if there's no way to tell if it's doing any good. Ditto for making a membrane-bound antibody. And why make a messenger protein first if there is nobody to give it a message, and nobody to receive the message if it did get one? We are led inexorably to the conclusion that even this greatly simplified clonal selection could not have come about in gradual steps.

Even at this simplified level, then, all three ingredients had to evolve simultaneously. Each of these three items—the fixed antibody, the messenger protein, and the loose antibodies—had to be produced by a separate historical event, perhaps by a coordinated series of mutations changing preexisting proteins that were doing other chores into the components of the antibody system. Darwin's small steps have become a series of wildly unlikely leaps. Yet our analysis overlooked many complexities: How does the cell switch from putting the extra oily piece on the membrane to not putting it on? The message system is fantastically more complicated then our simplified version. Ingestion of the protein, chopping it up, presenting it to the outside on an MHC protein, specific recognition of the MHC/fragment by a helper T cell, secretion of interleukin, binding of interleukin to the B cell, sending the signal that interleukin has bound into the nucleus—the prospect of devising a step-by-step pathway for the origin of the system is enough to make strong men blanch.

MIX AND MATCH

Factories float around in huge numbers, poised to deliver antibodies that can stick to an invader with virtually any shape. But how does the body make all those billions of differently shaped antibodies? It turns out that there is an elegant trick for making very many different antibodies without requiring enormous quantities of genetic material to code for the proteins. Over the next few pages I'll describe the system in some detail. Again, don't be concerned if the details quickly slip your mind; my purpose here is just to help you appreciate the complexity of the immune system.

☐ It took a fascinating discovery to lead scientists to puzzle out the full complexity of the immune system. The discovery started with a potentially cruel, but necessary, experiment. Just to see what would happen, chemists made some small molecules that do not occur in nature and then attached them to a protein. When the protein carrying the synthetic molecules was injected into a rabbit, the scientists were astonished to find that, yes, the rabbit made antibodies that bound tightly to the synthetic molecule. How could this be? Neither the rabbit nor its ancestors ever met the synthetic molecule, so how did it know how to make antibodies against it? Why should it recognize a molecule it had never seen before?

The puzzle of "antibody diversity" intrigued scientists studying immunology. Several ideas were floated as possible explanations. Proteins were known to be flexible molecules, and antibodies are proteins. So maybe when a new molecule is injected into the body an antibody wraps around it, molds itself to that shape, and then somehow freezes in that configuration. Or maybe, because defense is so vitally important, the DNA of organisms contains a vast number of genes for antibodies with many different shapes—enough to allow them to recognize things they hadn't seen yet. But such a huge number of antibodies would take up more than the available coding space in the DNA. So maybe there were only a few antibodies, and when the cell divided, maybe there was some way to make a lot of mutations in just the areas coding for the binding sites of the antibodies. That way each new B cell in the body could carry different mutations, coding for an antibody different from all other B cells. Or maybe the answer was a combination of these, or maybe it involved something completely new.

The answer to the problem of antibody diversity had to await an astonishing discovery: a gene coding for a protein didn't always have to be a continuous segment of DNA—it could be interrupted.[3] If we compare a gene to a sentence, it was as if a protein's code, "The quick brown fox jumps over the lazy dog" could be altered (without destroying the protein) to read "*The quick br*dkdjf bufjwkw nhru*own fox jumps over the la*pfeqmzda lfybnek sybagjufu *zy dog.*" The sensible DNA message was broken up by tracts of nonsense letters that somehow were not included in the protein. Further work showed that for most genes, corrections would be made—splicing out the nonsense—after an RNA copy is made of a DNA gene. Even with "interrupted" DNA, an edited and corrected message in RNA could be used by the cell's machinery to make the correct protein. Even more surprisingly, for antibody genes the DNA *itself* can also be spliced. In other words, DNA that is inherited can be altered. Amazing!

Splicing and rearrangement of DNA play a large role in explaining the great number of antibodies that the body can produce. The following is a brief description of work that has taken many investigators many years to accomplish; because of their efforts, the riddle of antibody diversity is solved.

At conception there are a number of gene pieces in the fertilized cell that contribute to making antibodies. The genes are arranged into clusters that I will simply call cluster 1, cluster 2, and so forth. In humans there are approximately 250 gene segments in cluster 1; a ways down the DNA from cluster 1 are ten gene segments that form cluster 2; further on down the DNA road are a group of six segments that comprise cluster 3; and down a piece from that are eight other gene segments that make up cluster 4. These are the players.

After the youngster grows a bit and sets his mind to getting born, one thing he wants to do is produce B cells. During the making of B cells, a funny thing happens: the DNA in the genome is rearranged, and some of it is thrown away. One segment from cluster 1 is picked out, apparently at random, and joined to one segment from cluster 2. The intervening DNA is cut out and discarded. Then a segment from cluster 3 is picked, again apparently at random, and joined to the cluster 1-2 segment.

The recombining of the segments is a little bit sloppy—not what you usually expect from a cell. Because of the sloppy procedure, the coding for a few amino acids (remember, amino acids are the build-ing blocks of proteins) can get added or lost. Once the cluster 1-2-3 segment is put together, the DNA rearrangement is over.[4] When it's time to make an antibody, the cell makes an RNA copy of the cluster 1-2-3 combination and adds to it an RNA copy of a segment from cluster 4. Now, finally, the regions that code for contiguous protein segments are themselves in a contiguous arrangement on the RNA.

How does this process explain antibody diversity? It turns out that portions of the segments from clusters 1, 2, and 3 form part of the binding site—the tips of the Y. Mixing and matching different segments from the three different clusters multiplies the number of binding sites with different shapes. For example, suppose that one segment from cluster 1 coded for a bump in the binding site, and another coded for a positive charge. And suppose that different seg-ments from cluster 2 coded for an oily patch, a negative charge, and a deep depression, respectively. Picking one segment randomly from cluster 1 and cluster 2, you could have six possible combinations: a bump next to an oily patch, negative charge, or deep depression; or a positive charge next to an oily patch, negative charge, or deep depression. (This is essentially the same principle whereby pulling three numbers out of a hat explains the diversity of a state lottery; picking just three numbers from 0 to 9 gives a total of one thousand possible combinations.) When making an antibody heavy chain, the cell can pick one of two hundred and fifty segments from cluster 1, one of ten from cluster 2, and one of six from cluster 3. Furthermore, the sloppiness during recombination "jiggles" the segments (by crowding another amino acid into the chain, or leaving one out); this effect adds another factor of about 100 to the diversity. By mixing and matching DNA segments you get $250 \times 10 \times 6 \times 100$, which is about a million different combinations of heavy-chain sequences. Similar processes produce about ten thousand different light-chain combinations. Matching one light-chain gene to one heavy-chain gene at random in each cell gives a grand total of ten thousand times one million, or ten billion combinations! The huge number of different antibodies provides so many different binding sites that it's almost certain at least one of them will bind almost any molecule—even synthetic ones. And all of this diversity comes from a total of just about four hundred different gene segments.

The cell has other tricks to tweak upward the number of possible antibodies. One trick happens after a foreign invasion. When a cell binds to foreign material, it receives a signal to replicate; during many rounds of replication the cell "intentionally" allows a very high level of mutation in just the variable regions of the heavy- and light-chain genes. This produces variations on a winning theme. Because the parent cell coded for an antibody that already was known to bind pretty well, mutating the sequence might produce a stronger binder. In fact, studies have shown that the antibodies produced by cells late in an infection bind much more tightly to foreign molecules than antibodies produced early in an infection. This "somatic hypermutation" adds another several orders of magnitude to the diversity of possible antibodies.

Remember the difference between B-cell factories and plasma factories? That oily piece of the Y that anchors the antibody in the B-cell membrane? For a plasma cell, when the RNA copy of the gene is made, the membrane segment is not copied. The segment is down-stream from the rest of the gene. The DNA can be likened to a message that says "*The quick b*rdkdjf bufjwkw nhru*own fox jumps over the la*pfeqmzda lfybnek sybagjufu *zy dog* kdjyf jdjkekiwif vmnd *and eats the* mnaiuw *rabbit.*" The final words can be left in or taken out, and the message still makes some sense. ❏

INCH BY INCH

An antibody-diversity system requires several components to work. The first, of course, is the genes themselves. The second is a signal identifying the beginning and end of gene segments. In modern organisms, each segment is flanked by specific signals that tell an enzyme to come along and join the parts together. This is like a sentence that reads "The quick br*cut here*[fjwkw]*cut here*own fox jumps over the la*cut here* [lfybnek sy]*cut here*zy dog"—as long as the beginning and ending are present, the cell knows to keep it together. The third component is the molecular machine that specifically recognizes the cutting signals and joins the pieces in the right order. In the absence of the machine, the parts never get cut out and joined. In the absence of the signals, it's like expecting a machine that's randomly cutting paper to make a paper doll. And, of course, in the absence of the message for the antibody itself, the other components would be pointless.

The need for minimal function reinforces the irreducible complexity of the system. Imagine you were adrift in a life raft on a stormy sea, and by chance a box floated by that contained an outboard motor. Your joy at the hope of deliverance would be short-lived if, after you affixed it to the boat, the outboard propeller turned at a rate of one revolution per day. Even if a complex system functions, the system is a failure if the level of performance is not up to snuff.

The problem of the origin of antibody diversity runs headlong into the requirement for minimal function. A primitive system with only one or a few antibody molecules would be like the propeller turning at one revolution per day: not sufficient to make a difference. (More to the point, it would be as if the FBI national identification database only contained two sets of fingerprints. Out of hundreds of thousands of criminals, the FBI could only hope to catch those two.) Because the likelihood is so small for the shape of one antibody being complementary to the shape of a threatening bacterium—perhaps one in a hundred thousand or so—an animal that spent energy making five or ten antibody genes would be wasting resources that could have been invested in leaving more progeny, or building a stronger skin, or making an enzyme for excretion that would degrade RNA. To do any good, an antibody-generating system would need to generate a very large number of antibodies from the start.

THE HIT MAN

Suppose it is a thousand years ago and you live in a large compound with a group of people. Because it is near the coast, you have to worry about Viking marauders. The compound is surrounded by a strong, high wooden fence; during a raid, pots of boiling oil are poured on folks trying to climb up ladders. One strange day a traveling wizard knocks on the compound door. Opening his pack, he offers to sell you a weapon from the future. He calls it a "gun." When the trigger is pulled, he says, the gun shoots a projectile in the direction you aim it. The gun is portable, and it could quickly be taken from one side of the compound to the other if the enemy sneakily shifted their attack. You and the other members of the compound pay the wizard two cows and four goats for the weapon.

Eventually there is a raid on your compound. Boiling oil flows freely, but the raiders have a battering ram. Hearing it whack the compound gate, you stride toward the gate confidently, gun in hand. Finally the gate is smashed and the raiders pour through, screaming and waving their battle axes. You aim the gun and fire at their leader. The projectile flies through the air and sticks to the Viking chieftain's nose. On the barrel of the gun, in letters you cannot read, is the inscription "Acme Toy Dart Gun." The chieftain stops, stares at you, and begins to grin as your smile dissolves. He and his friends rush at you; fortunately, you are reincarnated as a biochemist in the twentieth century.

Antibodies are like toy darts: they harm no one. Like a "Condemned" sign posted on an old house or an orange "X" painted on a tree to be removed, antibodies are only signals to other systems to destroy the marked object. It is surprising to think that after the body has gone to all the trouble to develop a complex system to generate antibody diversity, and after it has laboriously picked a few cells by the roundabout process of clonal selection, it is still virtually helpless against the onslaught of invaders.

❐ Much of the actual killing of foreign cells that are marked by antibodies is done by the "complement" system, which is called this because it complements the action of antibodies in getting rid of invaders. The pathway is remarkably complex (Figure 6–3); in many ways, it is similar to the blood-clotting cascade discussed in Chapter 4. It consists of about 20 kinds of proteins that form two related pathways, called the classical pathway and the alternative pathway. The classical pathway starts when a large aggregate of proteins, called C1, binds to an antibody that is itself bound to the surface of a foreign cell. It is crucial that the C1 complex recognize only bound antibody; if C1 attached itself to antibody that was floating around in the bloodstream, then all of the C1 would be sopped up and unavailable for action against enemies. Or, if C1 bound to the membrane-attached antibodies of B cells, it would initiate reactions that ultimately would end up killing good cells.

C1 is made up of 22 protein chains. These can be divided into three groups. The first is called C1q. It contains six copies of three

FIGURE 6–3

THE COMPLEMENT PATHWAY.

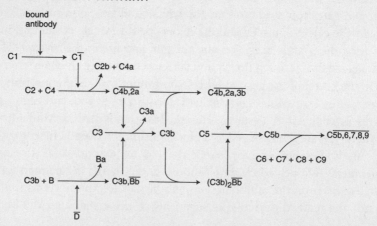

different types of proteins, for a total of 18. The other two groups
are called C1r and C1s. They both have two copies each of different proteins. The three different types of proteins in C1q all begin
with a special amino-acid sequence that resembles the sequence of
the skin protein collagen. The sequence allows the tails of the three
types of C1q proteins to wrap around each other like braids. This
arrangement holds one of each type of protein in a mini-complex.
The remainder of the protein chains then fold up into complex,
globular shapes at the top of the braid. Six of the minicomplexes
then come together. The six braids stick to each other lengthwise
to create a central stalk, out of which protrude six heads. Pictures
of C1q taken with an electron microscope show something resembling a hydra-headed monster. (Other people have likened it to a
bouquet of tulips, but I like more dramatic images.) The C1q heads
attach to the antibody–foreign cell complex. At least two of the
heads have to be attached before the pathway is initiated. Once
they stick, something in C1q changes, and the change in C1q
causes C1r and C1s to bind more tightly to C1q. When this happens C1r cuts itself (headline: Dog bites dog!) to give $C\overline{1r}$. ("Activated" proteins are designated by an upper bar over the number
and lower case letter.) $C\overline{1r}$ then is able to cut C1s to yield $C\overline{1s}$.

After C1s is cleaved, we still have a long way to go before the
work of destroying the invading cell is finished. The proteins of C1
are collectively called the "recognition unit." The next group of
proteins (named C2, C3, and C4) is called the "activation unit."
Unlike the recognition unit, the activation unit is not already together in one piece; it has to be assembled. The first step in forming the activation unit is the cleavage of C4 by $C\overline{1s}$. When C4 is
cut by $C\overline{1s}$, a very reactive group that was inside one piece (C4b) is
exposed to the surroundings. If the group is close to a membrane,
it can chemically react with it. The attachment of C4b is necessary
so the rest of the proteins in the activation unit can have an anchor
to hold them close to the invader. In contrast, if C4b is pointed in
the wrong direction or is floating around in solution, then the reactive group quickly decays without attaching to the correct membrane.

After C4b has attached itself to the target membrane, in association with $C\overline{1s}$ it cleaves C2 into two pieces. The larger piece, C2a,

remains stuck to C4b to yield $\overline{C4b,2a}$, also known as "C3 convertase." C3 convertase has to act quickly, or it falls apart and C2a floats away. If a molecule of C3 is in the vicinity, C3 convertase cleaves it into two pieces. C3b sticks to C3 convertase to form $\overline{C4b,2a,3b}$, which is also called "C5 convertase." The final reaction of the activation unit is the cleavage of C5 into two fragments.

At this point the system is finally ready to stick a knife in the invader. One of the pieces of C5 sticks to C6 and C7. This structure has the remarkable property of being able to insert itself into a cell membrane. $\overline{C5b,6,7}$ then binds to a molecule of C8 and a variable number (from one to eighteen) of molecules of C9 adds to it. The proteins, however, do not form an undifferentiated glob. Rather, they organize themselves into a tubular form that punches a hole in the membrane of the invading bacterial cell. Because the insides of cells are very concentrated solutions, osmotic pressure causes water to rush in. The in-rushing water swells the bacterial cell till it bursts.

There is an alternative pathway for the activation of the membrane-attack complex that can act quickly after infection, not needing to wait for the production of specific antibodies. In the alternative pathway a small amount of C3b, which apparently is produced continuously in low amounts, binds with a protein called factor B. C3b,B can then be cut by another protein, factor D, to give $\overline{C3b,Bb}$. This can now act as a C3 convertase. When more C3b is made, a second molecule of C3b can attach to yield $(C3b)_2\overline{Bb}$. Remarkably, this is now a C5 convertase, which produces C5b, which then goes on to start the formation of the membrane-attack complex in the way described above for the first pathway.

C3b is a dangerous protein to have floating around, since it can activate the destructive end of the complement pathway. In order to minimize random damage, two proteins (factors I and H), search out, stick to, and destroy C3b in solution. But if C3b is on the surface of a cell, then another protein (properdin), binds to and protects C3b from degradation so that it can do its job. How does C3b target foreign cells in the absence of antibodies? C3b is effective only if it sticks to the surface of a cell. The chemical reaction by which it does so goes faster in the presence of the molecules typically found on the surface of many bacteria and viruses. ☐

PROBLEMS, PROBLEMS

Like the blood-clotting pathway, the complement pathway is a cascade. Inevitably, in both cases one encounters the same problems trying to imagine their gradual production. It is not the final activity of a cascade that is the problem. The formation of a hole in a membrane does not necessarily require several different components; one killer protein could conceivably do the job. Nor does the formation of a protein aggregate, such as in blood clotting, necessarily require multiple components; under the right conditions, any protein will aggregate. (The particular shapes of the complement hole-complex and fibrin aggregate, however, are particularly suited to the jobs they do and need to be explained.) And as we saw in Chapter 4, a telephone pole by itself could bop Foghorn Leghorn.

It is the control systems that are the problem. At each control point both the regulatory protein and the masked protein that it activates have to be present from the beginning. If C5b were present, the rest of the cascade would immediately be touched off; but if C5 were present with nothing to activate it, then the whole pathway would always be shut off. If C3b were present, the rest of the cascade would immediately be touched off; but if C3 were present with nothing to activate it, then the whole pathway would always be shut off. Even if one imagines a much shortened pathway (where, say, C1s directly cuts C5), insertion of additional control points into the middle of the cascade runs into the same problem: the irreducible complexity of the switches.

◻ In addition to the generic problems of setting up a cascade, the complement pathway shares another problem with the blood-clotting cascade: attachment of proteins to membranes is crucial. Several clotting factors must first be modified to synthesize Gla residues so that they could stick to a membrane. In the complement pathway, both C3 and C4 have unusual, highly reactive internal groups that chemically attach to the membrane after the proteins are cleaved by other factors. These special features have to be available before the pathway is functional, adding a further severe barrier to their gradual development.

Numerous little features of the complement system are stumbling blocks to gradual development. Let's consider some subtle characteristics of just the C1 system. The three types of proteins in

C1q braid around each other, but do not braid with themselves. If they did, then the ratio of different types of chains in the complex would be changed, and there would be a much smaller chance of getting the real C1q complex with six copies of three different chains. If the binding of C1q to the antibody–foreign cell did not trigger C1r's self-scission, then the cascade would be stopped in its tracks. Conversely, if C1r cut itself before C1q bound to the antibody complex, then the cascade would be prematurely triggered. And so on. ❏

SISYPHUS WOULD SYMPATHIZE

The proper functioning of the immune system is a prerequisite for health. Major illnesses such as cancer and AIDS have either their cause or their cure, or both, in the vagaries of the system. Because of its impact on public health, the immune system is a subject of intense interest. Thousands of research laboratories around the world work on various aspects of the immune system. Their efforts have already saved many lives and promise to save many more in the future.

Although great strides have been made in understanding how the immune system works, we remain ignorant of how it came to be. None of the questions raised in this chapter has been answered by any of the thousands of scientists in the field; few have even asked the questions. A search of the immunological literature shows ongoing work in comparative immunology (the study of immune systems from various species). But that work, valuable though it is, does not address in molecular detail the question of how immune systems originated. Perhaps the best efforts at doing that so far have been in two short papers. The first, by Nobel laureate David Baltimore and two other prominent scientists, is tantalizingly entitled "Molecular Evolution of the Vertebrate Immune System." But it's hard to live up to such a title in just two pages. The authors point out that

> for any organism to have an immune system akin to that seen in mammals, the minimally required molecules are the antigen receptors (immunoglobulin and TCR), the antigen presentation molecules (MHC), and the gene rearranging proteins.[5]

(Immunoglobulins are antibodies. TCR molecules are akin to antibodies.) The authors then argue that sharks, which are very distantly related to mammals, appear to have all three components. But it's one thing to say an organism has a completed, functioning system, and another to say how the system developed. The authors certainly realize this. They note that

> immunoglobulin and TCR genes both require RAG proteins for rearrangement. On the other hand, RAG proteins require specific recombination signals to rearrange immunoglobulin and TCR genes.

(RAG is the component that rearranges the genes.) They make a valiant stab at accounting for the components, but in the end, it is a hop in the box with Calvin and Hobbes. The authors speculate that a gene from a bacterium might have luckily been transferred to an animal. Luckily, the protein coded by the gene could itself rearrange genes; and luckily, in the animal's DNA there were signals that were near antibody genes; and so on. In the final analysis the authors identify key problems with gradualistic evolution of the immune system, but their proffered solutions are really just a disguised shrug of their shoulders.

Another paper that gamely tries to account for a piece of the immune system is entitled "Evolution of the Complement System,"[6] Like the paper discussed above, it is very short and is a commentary article—in other words, not a research article. The authors make some imaginative guesses about what might come first and second, but inevitably they join Russell Doolittle in proposing unexplained proteins that are "unleashed" and "spring forth" ("At some point a critical gene fusion created a protease with a binding site for the primitive C3b"; "Evolution of the other alternative pathway components further improved the amplification and specificity"; and "C2, created by the duplication of the factor B gene, would then have allowed further divergence and specialization of the two pathways"). No quantitative calculations appear in the paper. Nor does an acknowledgment that gene duplications would not immediately make a new protein. Nor does any worry about a lack of controls to regulate the pathway. But then, it would be hard to fit those concerns in the four paragraphs of the paper that deal with molecular mechanisms.

There are other papers and books that discuss the evolution of the immune system.[7] Most of them, however, are at the level of cell biology and thus unconcerned with detailed molecular mechanisms, or else they are concerned simply with comparison of DNA or protein sequences. Comparing sequences might be a good way to study relatedness, but the results can't tell us anything about the mechanism that first produced the systems.

We can look high or we can look low, in books or in journals, but the result is the same. The scientific literature has no answers to the question of the origin of the immune system.

In this chapter I have looked at three features of the immune system—clonal selection, antibody diversity, and the complement system—and demonstrated that each individually poses massive challenges to a putative step-by-step evolution. But showing that the parts can't be built step by step only tells part of the story, because the parts interact with each other. Just as a car without steering, or a battery, or a carburetor isn't going to do you much good, an animal that has a clonal selection system won't get much benefit out of it if there is no way to generate antibody diversity. A large repertoire of antibodies won't do much good if there is no system to kill invaders. A system to kill invaders won't do much good if there's no way to identify them. At each step we are stopped not only by local system problems, but also by requirements of the integrated system.

We have looked at some positive features of the immune system, but there are also drawbacks to carrying around loaded weapons. You have to make sure you don't shoot yourself in the foot. The immune system has to discriminate between itself and the rest of the world. When, say, a bacterium invades, why does the body make antibodies against it but not against the red blood cells that are continually circulating in the bloodstream, or any of the other tissues that antibody cells constantly bump up against? When the body does make self-directed antibodies, it is generally a disaster. For example, people suffering from multiple sclerosis make antibodies that are directed against the insulation that surrounds nerves. That causes the immune system to destroy the insulation, exposing and short-circuiting nerves, and leading to paralysis. In juvenile diabetes, antibodies are made against the ß cells of the pancreas, leading to their destruction. The unfortunate person can no longer make insulin and usually dies unless insulin is supplied artificially. How the body acquires tolerance to its own tissues is still obscure, but whatever the mechanism, we know one thing: a system of self-toleration had to be present from the start of the immune system.

Diversity, recognition, destruction, toleration—all these and more interact with each other. Whichever way we turn, a gradualistic account of the immune system is blocked by multiple interwoven requirements. As scientists we yearn to understand how this magnificent mechanism came to be, but the complexity of the system dooms all Darwinian explanations to frustration. Sisyphus himself would pity us.

It is perhaps not surprising to discover unremitting complexity in such *Star Wars*–like machines as comprise the immune system. But what about humbler systems? What about the factories that manufacture the nuts and bolts out of which molecular machines are made? In a final evidence chapter I will examine the system that makes one of the "building blocks." We will see that complexity reaches down to the very bottom of the cell.

CHAPTER 7

ROAD KILL

LOOK BOTH WAYS

My family and I live about five miles from campus on one of the many beautiful mountains that grace Pennsylvania. The area, although close to town, is rural, with a thick forest wherever space has not yet been cleared for a house. Leading to our home is a narrow country road, winding this way and that as it makes its way up the mountain. As I drive to work in the morning or home at night I always see a few little animals crouching by the side of the road, ready to make a run for it. Whether they are taking a dare, trying to impress the opposite sex, or just anxious to get home, I do not know. But it is a dangerous game they play, and some pay the price.

Squirrels are the worst. Unlike more sensible animals, squirrels don't just cross over. While far away you can spot them sitting on one side of the road. As you get closer, they dash over to the other side, stop, reverse, and scramble back to the center. Closer and closer you get, and they're still in the road. Finally, as you drive by, they decide that your side is where they really want to be. Squirrels can fit under the car, so there's always hope as they disappear under the front end that you might see them in the rearview mirror, scurrying to safety. Sometimes they make it; sometimes they don't.

Groundhogs generally travel in a straight line across the road, making their position easy to anticipate, but you don't get much warning. Usually you're driving along, thinking about dinner, when all of a sudden a small, round shape waddles out of the darkness into your lane. At that point all you can do is grit your teeth and wait for the bump—unlike squirrels, groundhogs don't fit under the car. The next morning all that's left is a little stain on the road, other cars having obliterated the carcass. Nature red in tooth, claw, and tarmac.

Although traffic has picked up on the road lately, it's still pretty slow—one car every few minutes during the day, one every half hour at night. So most animals that cross the road easily make it to the other side. That's not true everywhere. The Schuylkill Expressway, the main highway into Philadelphia from the northwest, is eight or ten lanes wide in certain stretches. The volume of traffic can easily be thousands of times what it is on the road by my house. It would not be smart to bet on a groundhog starting from one side of the Schuylkill during rush hour getting to the other side.

Suppose you were a groundhog sitting by the side of a road several hundred times wider than the Schuylkill Expressway. There are a thousand lanes going east and a thousand lanes going west, each filled with trucks, sports cars, and minivans doing the speed limit. Your groundhog sweetheart is on the other side, inviting you to come over. You notice that the remains of your rivals in love are mostly in lane one, with some in lane two, and a few dotted out to lanes three and four; there are none beyond that. Furthermore, the romantic rule is that you must keep your eyes closed during the journey, trusting fate to deliver you safely to the other side. You see the chubby brown face of your sweetie smiling, the little whiskers wiggling, the soft eyes beckoning. You hear the eighteen-wheelers screaming. And all you can do is close your eyes and pray.

The example of groundhogs crossing a road illustrates a problem for gradualistic evolution. Up until this point in the book I have emphasized irreducible complexity—systems that require several components to function, and so are mammoth barriers to gradual evolution. I have discussed a number of examples; more can be seen just by paging through a biochemistry textbook. But some biochemical systems are not irreducibly complex. They do not necessarily require several parts to function, and there seem to be (at least at first blush) ways to assemble them step-by-step. Nonetheless, upon closer examination, nasty problems pop up. Supposedly smooth transitions turn out to be ephemeral when checked in the light of day. So even though some systems are not irreducibly complex, it does not necessarily mean that they have been put together in a Darwinistic manner. Like a groundhog trying to cross a thousand-lane highway, there is no absolute barrier to putting together some biochemical systems gradually. But the opportunities to go wrong are overwhelming.

THE BUILDING BLOCKS

The big molecules that do the work in the cell—proteins and nucleic acids—are polymers (that is, they are made of discrete units strung together in a row). The building blocks of proteins are amino acids, and the building blocks of nucleic acids are nucleotides. Much like a child's snap-lock beads, amino acids or nucleotides can be strung to give an almost infinite variety of different molecules. But where do the beads come from? Snap-lock beads are made in a factory; they aren't just found lying around in the woods. The factory makes the beads in specific shapes so that the little hole in one end is the right size for the knob sticking out of the other end. If the knob were too big, the beads could not be joined; if the holes were too big, the string of beads would fall apart. The manufacturer of snap-lock beads takes great care to mold them in the right shape and to use the right kind of plastic. The cell takes much care in manufacturing its building blocks, too.

DNA, the most famous of nucleic acids, is made up of four kinds of nucleotides: A, C, G, and T.[1] In this chapter I will talk mostly about the building block A. When the building block is not connected to a polymer, it can be in several forms, designated AMP, ADP, or ATP. The form that is first synthesized in the cell is AMP. Like snap-lock beads, AMP has to be made carefully. Most molecules in biological organisms are made of just a few different kinds of atoms, and AMP is no exception. It is comprised of five different kinds: ten carbons, eleven hydrogens, seven oxygens, four nitrogens, and one phosphorus.

I've used the analogy of snap-lock beads to convey how amino acids and nucleotides are put together into long chains. To understand how AMP is synthesized, let's think of something like Tinkertoys. For those readers who are unfamiliar with them, Tinkertoys have two kinds of pieces—a wooden wheel with holes drilled into the rim and center, and wooden sticks that have the same diameter as that of the holes. By pushing the sticks into the holes, you can connect several wheels. By using more sticks and wheels you can build up a whole network. The structures you can make from just those two types of pieces, from castles and cars to dollhouses and bridges, are limited only by your imagination. Atoms are like the pieces of a Tinkertoy set: the atoms are the wooden wheels, and the chemical bonds formed between atoms are the sticks. Like Tinkertoys, atoms can be put together to form many different shapes. A big difference is that the cell is a machine, however, so the mechanism to assemble the molecules of life must be automated. Imagine the complexity of a machine that could automatically assemble Tinkertoys into, say, the shape of a castle! The mechanism that the cell uses to make AMP is automated, and as expected, it is far from simple.

Atoms are almost always found in molecules; they're not lying free like tinker toy pieces. So to make a new molecule you generally have to take old molecules and join parts of them together. It's like taking a turret off of a Tinkertoy castle to use as a car body, using a propeller from a Tinkertoy airplane as a car wheel, etc. Similarly, new molecules are built up from pieces of old molecules. The molecules that are used to build up AMP all have rather long and tedious chemical names; I won't use them in the description unless I have to. Instead I'll just describe the molecules in words and give them innocuous names like "Intermediate III" and "Enzyme VII."

Figure 7–1 shows the molecules that are involved in the step-by-step synthesis. Most readers will probably find my description on the next several pages easier to follow by referring frequently to the figure. Don't worry, though—I'm not going to talk about any esoteric concepts; just who is connected to whom. The point is to appreciate the complexity of the system, to see the number of steps involved, to notice the specificity of the reacting components. The formation of biological molecules does not happen in some fuzzy-minded Calvin and Hobbes way; it requires specific, highly sophisticated molecular robots to get the job done. I urge you to skim along through the next two sections and marvel.

FIGURE 7–1

BIOSYNTHESIS OF AMP. THE FIGURE STARTS WITH INTERMEDIATE III. F
REPRESENTS THE "FOUNDATION"—RIBOSE-5-PHOSPHATE. WHITE
BOXES ARE NITROGEN ATOMS, BLACK ARE CARBON ATOMS, AND GRAY
ARE OXYGEN ATOMS. THE ATOMS ARE NUMBERED IN THE ORDER THEY
BECOME ATTACHED. ONLY ATOMS THAT WILL BE PART OF THE FINAL
PRODUCT ARE NUMBERED. ATOMS THAT BECOME ATTACHED BUT ARE
SUBSEQUENTLY REPLACED OR CUT OFF ARE MARKED WITH AN X.

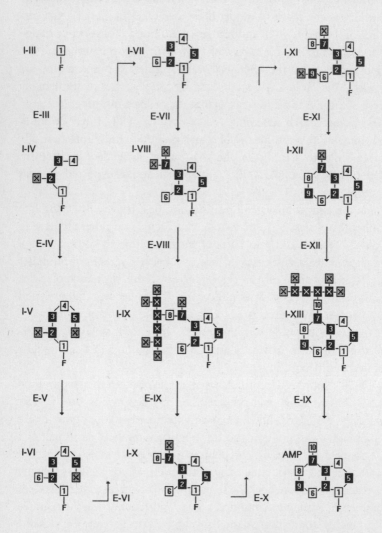

CONSTRUCTION STARTS

❐ To build a house you need energy. Sometimes the energy is just in the muscles of the workers, but sometimes it is in the gasoline that powers bulldozers or electricity that turns drills. The cell needs energy to make AMP. The cell's energy comes in discrete packages; I'll call them "energy pellets." Think of them as molecular candy bars, to provide energy for muscles, or gallon cans of gasoline, to power machines. There are several different types of energy pellets, including ATP and GTP. Don't worry about what they look like or how they work; I'll just note at which steps we need them.

The first two steps in the synthesis of AMP aren't shown in Figure 7–1—they happen offstage. Just as the building of a house starts with the foundation, so does the synthesis of AMP. The foundation is a complicated molecule whose synthesis I will not discuss. It consists of a ring of atoms: four carbons and one oxygen. To three of the ring carbons are attached oxygen atoms. To the fourth carbon in the ring is attached another carbon, to which is hooked an oxygen, to which is attached a phosphorus with three oxygens. In the first step of the synthesis of AMP a group consisting of two atoms of phosphorus and six atoms of oxygen is transferred by Enzyme I, en masse, to one of the oxygens of the foundation to make Intermediate II. This requires an energy pellet of ATP. Intermediate II is used by the body as the starting point for making several different molecules, including AMP.

In the next step Enzyme II takes a nitrogen atom from the amino acid glutamine and places it on a ring carbon to give Intermediate III. In the same step the phosphorus/oxygen group that was attached in the last step is kicked off. This is the point at which Figure 7–1 takes up the story. To make the figure easier to follow, I will just represent the foundation by the letter F. So at this point in Figure 1 we see a a nitrogen atom attached to a letter F.[2] Nitrogen atoms are colored white in the figure, carbons are black, and oxygens are gray. The atoms that will end up in the final product (AMP) are numbered according to the order in which they are attached. Atoms that won't end up in AMP are marked with an "X."

Under the guidance of Enzyme III, an amino acid called glycine (consisting of a nitrogen atom that is attached to a carbon, which is attached to another carbon attached to two oxygens) glides in and hooks on to the nitrogen of Intermediate III through one of its carbon atoms. This uses an energy pellet of ATP. In the process one of the two oxygens originally attached to carbon #2 is kicked out. At this point the molecule looks like the foundation has a tail waving in the breeze. The finished product, AMP, is going to look very different: a couple of stiff, fused rings attached to the foundation. In order to get there from where we are now, the molecule has to be chemically prepared in the right order.

In the next step a molecule of formic acid (actually the related ion, formate), consisting of two atoms of oxygen attached to an atom of carbon, is stuck onto nitrogen #4 of Intermediate IV to make Intermediate V. In the process one of the formate oxygens is kicked out. Ordinarily formate is unreactive, so getting it to hook onto other molecules requires some preparation. A biochemistry textbook emphasizes the problem:

> Formate . . . is quite unreactive under physiological conditions and must be activated to serve as an efficient formylating agent. . . . The fundamental importance of [THF] is to maintain formaldehyde and formate in chemically poised states, not so reactive as to pose toxic threats to the cell but available for essential processes by specific enzymatic action.

Thankfully, as the quote points out, formate is not just floating around in solution. It is first attached to a vitamin called THF, a cousin of the B vitamin folic acid (don't even ask how the vitamin is synthesized). When it is attached by an enzyme to the vitamin (in a reaction requiring an energy pellet of ATP), formate is revved up and made ready for action. The THF-formate complex, however, would not join up with Intermediate IV to give Intermediate V unless directed to do so by Enzyme IV; it would float away in the cell until it reacted with something else or decayed, and that would mess up our synthesis of AMP. That doesn't happen, however, because the enzyme guides the reaction to the correct products.

The next step is to replace the oxygen atom that is hooked onto carbon #2 of Intermediate V with a nitrogen atom. This can be done chemically by exposing the molecule to ammonia—but you can't just throw ammonia into the cell, because it would react willy-nilly with a lot of things that it shouldn't react with. So part of an amino acid is used to donate the nitrogen atom that's needed. The amino acid glutamine, under the watchful eyes of Enzyme V, sidles up to Intermediate V so that the nitrogen of the amino acid is close to the first oxygen of Intermediate V. Through the catalytic wizardry that enzymes are famous for, the nitrogen hops off the amino acid, the oxygen is kicked out of Intermediate V, and the nitrogen takes its place to make Intermediate VI. This step uses an energy pellet of ATP. ❐

RING AROUND THE ROSIE

❐ The next step in building ourselves a molecule of AMP is in some ways like the last step. Again we're going to take a nitrogen atom and use it to replace an oxygen atom that's attached to a carbon, and again this step uses an energy pellet of ATP. But this time we don't have to bring in a nitrogen from the outside. Instead we'll use nitrogen #1, which is already in our molecule. The first nitrogen that was put on the foundation—the one that kicked out the phosphorus/oxygen group a number of steps ago—now comes into play. It takes the place of the oxygen atom that is last in the chain. But unlike the nitrogen that came from the amino acid in the previous step, this nitrogen doesn't break any of its bonds with other atoms. It just makes a new one, as seen in Intermediate VII. An interesting thing about this arrangement is that it now makes a ring of atoms; the ring has five members, with two groups sticking off of it. The first group is nitrogen #6, which was introduced in the last step, and the second group is the foundation.

When you shake a can of soda and open the lid, usually you get soaked by a spray of liquid. The spray is powered by the sudden release of carbon dioxide gas that had been dissolved in the liquid. Some carbon dioxide is also dissolved in cellular fluid (although an animal usually doesn't fizz when shaken) and can be used in biochemical reactions. That's good, because the next step in the synthesis of AMP needs carbon dioxide. In the reaction the gas molecule (actually its water-logged counterpart, bicarbonate) is placed by Enzyme VII onto carbon #3 to make Intermediate VIII. An energy pellet of ATP powers this step.[4]

And now it's time for another ammonia to be added. This step will also use an ATP energy pellet. Like the last time ammonia was added, it won't be found floating around free in solution (like the carbon dioxide was); it will be donated by an amino acid. But this time it will be the amino acid called aspartic acid. And, in another twist, the nitrogen does not leave the amino acid when it reacts with Intermediate VIII: we get the nitrogen we want, but also an ugly extra chain of atoms dangling off the end of Intermediate IX. Enzyme IX removes the unwanted appendage, sawing off only the extraneous part.

The result, Intermediate X, is a half-built molecule. Another molecule of activated formate—again hooked on to a vitamin—is attached to nitrogen #6 of Intermediate X to give Intermediate XI. In the next step, Enzyme XI directs nitrogen #8 to kick out the oxygen of the formate that was just attached and to make a bond to carbon #9; this gives Intermediate XII. Because the reacting nitrogen does not break its bond with the carbon to which it was initially attached, the reaction forms another ring. The two fused rings of Intermediate XII are rigid, not floppy like the chains of atoms that preceded ring formation. The formation of the six-member ring in this step is similar to the formation of the five-member ring several steps ago, and the reaction of formate in the last step is chemically similar to the previous addition of formate. But even though the two sets of steps are similar, they are catalyzed by two different sets of enzymes. This is necessary because the shape of the molecule has changed during synthesis, and enzymes are frequently sensitive to shape changes.

Intermediate XII is a nucleotide called IMP, which is used in some biomolecules (for example, one special type of RNA that helps to make protein contains a little bit of IMP). To make AMP from IMP requires a couple of different steps, which are shown in Figure 7–1. In a step reminiscent of an earlier one, Enzyme XII attaches a molecule of the amino acid aspartic acid to the six-membered ring, kicking out the oxygen atom with the nitrogen atom of the incoming molecule. This gives Intermediate XIII. The reaction uses an energy pellet, but not ATP; instead, for reasons I will discuss later, it uses GTP. Again, as happened last time that aspartic acid was attached, this leaves us with an ugly, detrimental appendage. Enzyme IX comes back (the only enzyme to be used twice in the pathway) to saw off the unnecessary part and leave behind the required nitrogen atom.

Finally we have AMP—one of the 'building blocks' of nucleic acids. ☐

GETTING THERE

I assume I've lost most readers in the labyrinth by now, so let me play accountant and summarize the biosynthesis of AMP. The synthesis takes thirteen steps and involves twelve enzymes; one of the enzymes, IX, catalyzes two steps. Besides the foundation molecule, ribose-5-phosphate, the synthesis requires five molecules of ATP to provide the energy to drive chemical reactions at different steps, one molecule of GTP, one molecule of carbon dioxide, two molecules of glutamine to donate nitrogen atoms at different steps, a molecule of glycine, two formyl groups from THF at separate steps, and two molecules of aspartic acid to donate nitrogen atoms at another two steps. Additionally, at two separate steps the remains of aspartic acid molecules have to be cut off, and at two separate steps parts of the growing molecule have to be reacted with each other to close the two rings. All thirteen steps occur to produce just one kind of molecule. The precursor molecules along the synthetic pathway—Intermediates III to XI—play no independent role; they are used for nothing but to make AMP or GMP.

All roads lead to Rome, it is said, and similarly there are many ways to synthesize AMP. A book for chemists that I have on my shelf lists eight different ways to make adenine (which is the top part of AMP, without the foundation);[5] the remainder of the molecule can be put together in a variety of ways also. Chemists who want to synthesize adenine, however, use completely different routes from that used by cells. Because they involve reactions in oily liquids at extremes of acidity, these conditions would cause the quick demise of any known organism.

In the early 1960s scientists who were interested in the origin of life discovered an interesting way to synthesize adenine.[6] They saw that the simple molecules hydrogen cyanide and ammonia—which are thought to have been plentiful in the early days of earth—will form adenine under the right conditions. The ease of the reaction so impressed Stanley Miller that he called it "the rock of the faith" for origin-of-life researchers.[7] But there's a problem lurking in the background: hydrogen cyanide and ammonia are not used in the biosynthesis of AMP. But even if they were on the ancient earth, and even if that had something to do with the origin of life (which is problematic on a number of other grounds), the synthesis of adenine from simple molecules in a chemist's flask gives us absolutely no information about how the route for making the molecule first developed in the cell.

Stanley Miller was impressed by the ease of synthesis of adenine from simple molecules, but the cell eschews simple synthesis. In fact, if we dissolved in water (using the formal chemical names) ribose-5-phosphate, glutamine, aspartic acid, glycine, N^{10}-formyl-THF, carbon dioxide, and energy packets of ATP and GTP—all the small molecules that are used by the cell to build AMP—and let them sit for a long time (say, a thousand or a million years) we would not get any AMP.[8] If Stanley Miller mixed these chemicals hoping for another rock of the faith, he would be quite disappointed.

Shoes might be all we need to get to Rome from Milan. But we will need more than shoes to get to Rome from Sicily; we will need a boat. And to get to Rome from Mars, we need very high-tech equipment indeed. To make AMP from the ingredients that the cell uses we also need very high-tech equipment: the enzymes that catalyze the reactions of the pathway. In the absence of the enzymes, AMP is simply not made by the reactions shown in Figure 7–1. The point is that even if adenine or AMP can be made by simple pathways, those pathways are no more precursors to the biological route of synthesis than shoes are precursors to rocket ships.

$$A \rightarrow B \rightarrow C \rightarrow D$$

Consider a metabolic pathway where compound A is transformed into compound D by way of intermediates B and C. Could the pathway have evolved gradually? It depends. If A, B and C are useful compounds for the cell, and if neither B, C, nor D are essential from the beginning, then perhaps a slow development is possible. In that instance we can imagine a cell that made A leisurely mutating so that, serendipitously, compound B was produced. If it did no harm, then perhaps over time the cell would find a use for compound B. And then perhaps the scenario could be repeated. A random mutation causes the cell to produce some C from B, a use is found for C, and so on.

However, suppose D is necessary from the beginning. AMP is required for life on earth: it is used to make DNA and RNA, as well as a number of other critical molecules. There may be some way to construct a living system that does not require AMP, but if there is, no one has a clue how to do so. The problem for Darwinian evolution is this: if only the end product of a complicated biosynthetic pathway is used in the cell, how did the pathway evolve in steps? If A, B, and C have no use other than as precursors to D, what advantage is there to an organism to make just A? Or, if it makes A, to make B? If a cell needs AMP, what good will it do to just make Intermediate III, or IV, or V? On their face, metabolic pathways where intermediates are not useful present severe challenges to a Darwinian scheme of evolution. This goes in spades for something like AMP, because the cell has no choice. AMP is required for life. Either it immediately has a way to produce or obtain AMP, or the cell is dead.

A few textbooks mention this problem. The typical explanation is economically expressed by Thomas Creighton:

> How might the biochemical complexity of metabolic pathways have evolved? In the case of the biosynthetic pathways that produce the building blocks of amino acids, nucleotides, sugars, and so forth, it is likely that these building blocks were originally present in the primordial soup and were used directly. As organisms increased in number, however, these constituents would have become scarce. Any organism that could produce one of them from some unused component of the primordial soup, using a newly evolved enzyme, would have had a selective advantage. Once the availability of that component became limiting, there would have been selection for any organism that could produce it from some other component of the primordial soup. According to this scenario, the enzymes of metabolic pathways would have evolved in a sequence opposite to the one they have in the modern pathway.[9]

Simply put, Creighton says that if we find a reaction pathway in a modern organism that goes A→B→C→D, then D was available in the primordial soup—synthesized by simple chemical precursors without benefit of enzymes. As the supply of D ran low, some organism would "learn" to make D from C. As C ran out, it would make C from B. When famine threatened again, it would learn to make B from A, and so on. The same scheme is described in *Molecular Biology of the Cell*, a popular text written by Nobel laureate James Watson, president of the National Academy of Sciences, Bruce Alberts, and several other coauthors. We are told in a figure legend that the primordial cell

> is provided with a supply of related substances (A, B, C, and D) produced by prebiotic synthesis. One of these, substance D, is metabolically useful. As the cell exhausts the available supply of D, a selective advantage is obtained by the evolution of a new enzyme that is able to produce D from the closely related substance C.[10]

Yes, everybody agrees that, if you run out of D, the thing to do is to make it from C. And of course, it should be a simple matter to convert B to C. After all, they're right next to each other in the alphabet. And where do we get A, B, and the rest? From the primordial alphabet soup, of course.

The fact is that no one ever puts real chemical names on any of the mythical letters in the A→B→C→D story. In the textbooks mentioned above, the cartoon explanations are not developed any further, even though the books are used to teach Ph.D. students who could easily follow detailed explanations. It is certainly no trouble to imagine that the primordial soup might have some C floating around which could easily be converted to D; Calvin and Hobbes could imagine that without any difficulty whatsoever. It is, however, much more difficult to believe there was much adenylosuccinate (Intermediate XIII) to be converted to AMP. And it is even harder to believe that carboxyaminoimidazole ribotide (Intermediate VIII) was sitting around waiting to be converted to 5-aminoimidazole-4-(N-succinylocarboxamide) ribotide (Intermediate IX). It is difficult to believe because, when you put real names on the chemicals, then you have to come up with a real chemical reaction that could make them. No one has done that.

The problems with the A→B→C→D theory are legion. Let's look at a few of the more prominent ones. First, except for Intermediate X, prebiotic synthesis experiments have yielded none of the intermediates in the biosynthesis of AMP.[11] Although adenine can be made by reacting ammonia and hydrogen cyanide, biochemical precursors to adenine can not. Second, there are good chemical reasons to think that intermediates in the biochemical pathway can't be made except under the careful guidance of enzymes. For example, if the right enzymes were not available to steer the reactions to Intermediates V and XI, formate would more likely react in nonproductive ways than in the ways required to make AMP. Note that those enzymes would have to be available before enzymes for the succeeding steps could be developed, else the later enzymes would have nothing to work on. Furthermore, the steps that require energy pellets have to be carefully guided so that the energy isn't squandered doing something useless. For example, the energy of gasoline can make a car move because it is channeled in the right way by a complex machine; burning gasoline in a pool under the car doesn't move it at all. Unless there was an enzyme guiding the use of the ATP energy pellet, the energy would be squandered. Notice once more that the enzymes needed to guide these steps would be required before the organism would have the chemical that is made in the next step of the pathway.

A third problem with the A→B→C→D story is that some of the intermediates in the pathway are chemically unstable. So even if, against all hope, they were made in an undirected prebiotic reaction, they would either quickly fall apart or quickly react in the wrong way; again they would not be available to continue the pathway. Other reasons could be advanced against the A→B→C→D story, but this will suffice.

THEN AND NOW

A few years ago I read *The Closing of the American Mind* by Allan Bloom. I was startled by his claim that many modern American ideas actually have their roots in old European philosophies. In particular I was surprised that the song "Mack the Knife" was a translation of a German song, "Mackie Messer," whose inspiration Bloom traces to a murderer's "joy of the knife" that Nietzsche describes in *Thus Spake Zarathrusta*.[12] Most of us like to think that our ideas are our own—or at least, if they were proposed by someone else, that we only agreed to them after conscious review and assent. It's unnerving to think, as Bloom maintained, that many of our important ideas about the way the world works were simply picked up unreflectively from the cultural milieu in which we found ourselves.

The A→B→C→D story is an old idea that has been passed on unreflectively. It was first proposed in 1945 by N. H. Horowitz in the *Proceedings of the National Academy of Sciences*. Horowitz sees the problem:

> Since natural selection cannot preserve nonfunctional characters, the most obvious implication of the facts would seem to be that a stepwise evolution of biosyntheses, by the selection of a single gene mutation at a time, is impossible.[13]

But there is hope:

> In essence, the proposed hypothesis states that the evolution of the basic syntheses proceeded in a stepwise manner, involving one mutation at a time, but that the order of attainment of individual steps has been in the reverse direction from that in which the synthesis proceeds, i.e., the last step in the chain was the first to be acquired in the course of evolution, the penultimate step next, and so on. This process requires for its operation a special kind of chemical environment; namely, one in which end products and potential intermediates are available. Postponing for the moment the question of how such an environment originated, consider the operation of the proposed mechanism. The species is at the outset assumed to (require) an essential organic molecule, D. . . . As a result of biological activity, the amount of available D is depleted to a point where it limits the further growth of the species. At this point, a marked selective advantage will be enjoyed by mutants which are able to carry out the reaction B + C = D. . . . In time B may become limiting for the species, necessitating its synthesis from other substances.[14]

Here is the source for the explanation of the development of biochemical pathways given by modern textbooks. But what was the state of science in Horowitz's day? In 1945, when his article appeared, the nature of a gene was unknown, as were the structures of nucleic acids and proteins. No experiments had yet been done to see if the "special kind of chemical environment" Horowitz postulated was possible. In the intervening years biochemistry has progressed tremendously, but no advance encourages his hypothesis. The structures of genes and proteins are known to be much more complicated than thought in Horowitz's day. There are good chemical reasons for thinking that the intermediates in AMP synthesis would not be available outside of a living cell, and no experiment has shown otherwise. The "moment" for which Horowitz postponed "the question of how such an environment originated" has now stretched past fifty years. Despite the manifest difficulties, the old story is repeated in textbooks as if it were as obvious as the nose on your face; the progress of five decades can't put a dent in received wisdom. Reading modern texts, you can almost hear the haunting strains of "Mack the Knife."

Although textbooks carry the standard idea, some people are restless. Nobel laureate Christian de Duve, in his book *Blueprint for a Cell,* expresses skepticism of the importance of the hydrogen cyanide/ammonia pathway. Instead he proposes that AMP arose through "protometabolic pathways" in which a lot of little proteins just happened to have the ability to make a lot of different chemicals, some of which were intermediates in the AMP pathway. To illustrate his theory he has a figure in which arrows point from the words *abiotic syntheses* to the letters A, B, C, and D. But, breaking new ground, he has arrows pointing from A, B, C, and D to M, N, S, T, and W, and from there to P, O, Q, R, and U. Beside each of the arrows he has written *Cat* (as an abbreviation for "catalyst") to show how the letters originated, but that is no explanation: the only "evidence" for the scheme is the figure! Nowhere does he or any other researcher attach names of real chemicals to the mythical letters. Origin-of-life workers have never demonstrated that the intermediates in the synthesis of AMP either would have or even could have existed in a prebiotic soup, let alone sophisticated enzymes for interconverting the intermediates. There is no evidence that the letters exist anywhere outside of de Duve's mind.

Another restless scientist is Stuart Kauffman of the Santa Fe Institute. The complexity of the metabolism of living organisms makes him doubt that a step-by-step approach would work:

> In order to function at all, a metabolism must minimally be a connected series of catalyzed transformations leading from food to needed products. Conversely, however, without the connected web to maintain the flow of energy and products, how could there have been a living entity to evolve connected metabolic pathways?[15]

To answer his question he proposes, in very mathematical terms, something similar to what de Duve toyed with: a complex mixture in which some chemicals happen to be transformed into other chemicals that are transformed into still others, and somehow this forms a self-sustaining network. It is clear from his writings that Kauffman is a very smart guy, but the connection of his mathematics to chemistry is tenuous at best. Kauffman discusses his ideas in a chapter entitled "The Origin of a Connected Metabolism," but if you read the chapter from start to finish *you will not find the name of a single chemical*—no AMP, no aspartic acid, no nothing. In fact, if you scan the entire subject index of the book, you will not find a chemical name there either. John Maynard Smith, Kauffman's old mentor, has accused him of practicing "fact-free science."[16] That is a harsh accusation, but the complete lack of chemical details in his book appears to justify the criticism.

Kauffman and de Duve identify a real problem for gradualistic evolution. The solutions they propose, however, are merely variations on Horowitz's old idea. Instead of A→B→C→D, they simply propose A→B→C→D times one hundred. Worse, as the number of imaginary letters increases, the tendency is to get further and further away from real chemistry and to get trapped in the mental world of mathematics.

TOO MUCH OF A GOOD THING

Every child at one time or another hears the tale of King Midas. The greedy king loved gold more than anything, or so he thought. When he was first given the magical gift of turning anything to gold by his touch, he was delighted. Old vases, worthless stones, used clothing, all became beautiful and priceless by mere contact with him. However, storm clouds could be sighted when Midas touched already-beautiful flowers, which then lost their fragrance. He knew he was in deep trouble when the food he tried to eat turned to gold. Finally, folly led to grief when his daughter, little Marygold, hugged her father and turned into a golden statue.

The story of King Midas teaches some obvious lessons: don't be greedy, love is worth more than money, and so forth. But there is another, less obvious lesson about the importance of regulation. It is not enough to have a machine or process (magical or otherwise) that does something; you have to be able to turn it on or off as needed. If the king had wished for the golden touch *and* the ability to switch it on or off when he wanted, he could have transmuted a few rocks into gold nuggets but not zap his daughter. He could turn the plates to gold, but not the food.

The need for regulation is obvious for machines we use in our daily lives. A chain saw that couldn't be turned off would be quite a hazard, and a car with no brakes and no neutral gear would be of little use. Biochemical systems are also machines we use in our daily lives (whether we think of them or not), and so they too have to be regulated. To illustrate this, let's spend the next three paragraphs looking at the ways in which the synthesis of AMP is regulated (outlined in Figure 7–2).

❐ Enzyme I requires an ATP energy pellet to transform ribose-5-phosphate (the foundation) into Intermediate II. The enzyme has

FIGURE 7–2

REGULATION OF THE AMP PATHWAY. HEAVY WHITE ARROWS INDICATE COMPOUNDS THAT SLOW DOWN SYNTHESIS; HEAVY BLACK ARROWS INDICATE COMPOUNDS THAT SPEED UP SYNTHESIS.

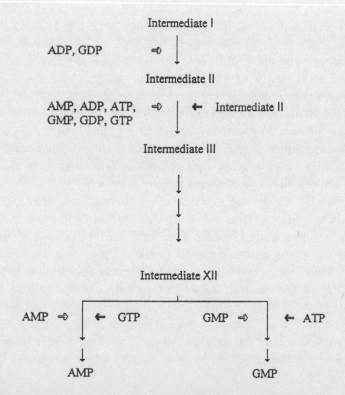

an area on its surface that can bind either ADP or GDP when there is an excess of those chemicals in the cell. The binding of ADP or GDP acts as a valve, decreasing the activity of the enzyme and slowing the synthesis of AMP. This makes good physiological sense: since ADP is the remains of a spent ATP (like a bullet shell after a gun has been fired), high concentrations of ADP in the cell means that the concentration of ATP, the cellular energy pellet, is low. Instead of making AMP, Intermediate I is then used as fuel to produce more ATP.

Commonly in biochemistry, the first enzyme that irrevocably starts a molecule down a particular metabolic pathway is highly regulated. The AMP pathway is no exception. Although Intermediate II can be used for other things, once it is transformed into Intermediate III the molecule is inevitably swept on to either AMP or GMP by the other enzymes of the pathway. So the enzyme that catalyzes the critical reaction (Enzyme II) is also regulated. Enzyme II, in addition to binding sites for the reacting molecules, has two other binding sites on its surface: one that will hold either AMP, ADP, or ATP, and a second site that will hold either GMP, GDP, or GTP. If one site is filled, the enzyme works more slowly; if both sites are filled, it works more slowly yet. Furthermore, in addition to the site where reaction takes place, Enzyme II contains another site that binds Intermediate II, itself a reactant. Binding of Intermediate II to the second site makes the enzyme work faster. Again this makes physiological sense: if there is so much Intermediate II around that it binds to both sites of the enzyme, then the cell is behind in its synthetic work and needs to process Intermediate II more quickly.

Synthesis is regulated at several other places as well. After IMP is made the pathway splits to build either AMP or GMP. Enzyme XII, which catalyzes the first step from IMP to AMP, is itself slowed down by excess amounts of AMP. Similarly, the catalysis of the first step from IMP to GMP is inhibited by excess GMP. (Unlike King Midas, the enzymes can tell when they have too much of a good thing.) Finally, Enzyme XII uses GTP as an energy pellet because, if a lot of GTP is around, more "A" nucleotides (AMP, ADP, and ATP) are needed to keep the supply in balance. The final step in the synthesis of GMP uses ATP as an energy source for similar reasons. ❐

REGULATORY FAILURE

When the regulation of metabolism fails, the result is illness or death. An example is diabetes; the uptake of sugar into cells is slowed, even though sugar molecules that manage to get into cells are otherwise metabolized normally. A disease, much less common than diabetes, that results from a failure to regulate AMP synthesis is called Lesch-Nyhan syndrome. In Lesch-Nyhan syndrome an enzyme needed to re-cycle used nucleotides from degraded DNA or RNA is missing or inac-tive; this indirectly causes Intermediate II to accumulate. Unfortunately, as mentioned above, Intermediate II stimulates Enzyme II, which in turn increases the synthesis of AMP and GMP. The in-creased synthesis leads to the production of excess uric acid (the breakdown product of AMP and GMP), which comes out of solution and crystallizes. Random deposits of uric acid crystals can disrupt nor-mal body functions, as they do in gout. In Lesch-Nyhan syndrome, however, the consequences are more severe. They include mental re-tardation and a compulsion toward self-mutilation—the patient bites his own lips and fingers.

The regulation of AMP biosynthesis is a good example of the intri-cate mechanisms needed to keep the supply of biomolecules at the right level: not too much, not too little, and in the right ratio with re-lated molecules. The problem for Darwinian gradualism is that cells would have no reason to develop regulatory mechanisms before the appearance of a new catalyst. But the appearance of a new, unregu-lated pathway, far from being a boon, would look like a genetic disease to the organism. This goes in spades for fragile ancient cells, putatively developing step by step, that would have little room for error. Cells would be crushed between the Scylla of unavailability and the Charyb-dis of regulation.

No one has a clue how the AMP pathway developed. Although a few researchers have observed that the pathway itself presents a severe challenge to gradualism, no one has written about the obstacle posed by the need to regulate a cell's metabolic pathway immediately at its inception. Small wonder—no one wants to write about road kill.

In the distant past, a cell gazes across the wide highway. On the other side is a brand new metabolic pathway. The chemical trucks, buses, station wagons, and motorcycles zoom by without noticing the little fellow. In the first lane, marked "intermediates not found in soup," he sees the remains of most earlier cells that heard the siren call. There are a few cellular remains in lane two, marked "guiding mechanism required." One or two are in the third lane, "instability of intermediates." There are no cell bodies in lane 4, "regulation"; none made it that far. The other side is very distant indeed.

STRICT CONSTRUCTION

The Ninth Amendment to the Constitution of the United States stipulates that "The enumeration in the Constitution, of certain rights, shall not be construed to deny or disparage others retained by the people." That's a handy way to say that a short document can't hope to cover all bases, so nothing is implied about things that have not been discussed. I would like to make a similar disclaimer about this book. In Chapters 3 to 6 I discussed several irreducibly complex biochemical systems, going into a lot of detail to show why they could not be formed in a gradualistic manner. The detail was necessary so that the reader could understand exactly what the problems are. Because I spent a lot of time on those systems I didn't have time to get on to other biochemical systems, but this does not imply that they are not also problems for Darwinism. Other examples of irreducible complexity abound, including aspects of DNA replication, electron transport, telomere synthesis, photosynthesis, transcription regulation, and more. The reader is encouraged to borrow a biochemistry textbook from the library and see how many problems for gradualism he or she can spot.

This chapter was somewhat different. In this chapter I wanted to show that it is not only irreducibly complex systems that are a problem for Darwinism. Even systems that at first glance appear amenable to a gradualistic approach turn out to be major headaches on closer inspection—or when the experimental results roll in—with no reason to expect they will be solved within a Darwinian framework.

The idea originally offered by Horowitz was a good one in its day. It could have worked; it might have been true. Certainly if a complex metabolic pathway ever arose gradually, the scheme Horowitz outlined must have been the way it happened. But as the years passed and science advanced, the prerequisites for his scheme crumbled. If there is a detailed Darwinian explanation for the production of AMP out there, no one knows what it is. Hard-nosed chemists have begun to drown their frustrations in mathematics.

AMP is not the only metabolic dilemma for Darwin. The biosynthesis of the larger amino acids, lipids, vitamins, heme, and more run into the same problems, and there are difficulties beyond metabolism. But the other problems will not concern us here. I will now turn my attention away from biochemistry per se and focus on other issues. The scientific obstacles discussed in the last five chapters will serve as stark examples of the mountains and chasms that block a Darwinian explanation of life.

PART III

WHAT DOES THE BOX TELL US?

CHAPTER 8

PUBLISH OR PERISH

THE JOURNAL OF MOLECULAR EVOLUTION

In Chapters 3 through 7, I showed that no one has explained the origin of the complex biochemical systems I discussed. There are tens of thousands of scientists in the United States, however, who are interested in the molecular basis of life. Most of them spend their time in the hard work of isolating proteins, analyzing structures, and sorting out the details of the ways that Lilliputian things work. Nonetheless, some scientists are interested in evolution and have published a large amount of work in the professional literature. If complex biochemical systems are unexplained, what type of biochemical work has been published under the heading of "evolution"? In this chapter you will see what has been studied—and what hasn't.

When the molecular basis of life was discovered, evolutionary thought began to be applied to molecules. As the number of professional research papers in this area expanded, a specialty journal, the *Journal of Molecular Evolution*, was set up. Established in 1971, *JME* is devoted exclusively to research aimed at explaining how life at the molecular level came to be. It is run by prominent figures in the field. Among the more than fifty people who make up the editorial staff and board, are about a dozen members of the National Academy of Sciences. The editor is a man named Emile Zuckerkandl, who (along with Linus Pauling) first proposed that differences in the amino acid sequences of similar proteins from different species could be used to determine the time at which the species last shared a common ancestor.

Each monthly issue of *JME* contains about ten scientific papers on various aspects of molecular evolution. Ten papers per month means about a hundred papers per year, and about a thousand papers per decade. A survey of a thousand papers in a particular area can give you a pretty good idea of what problems have been solved, what problems are being addressed, and what problems are being ignored. A look back over the last decade shows that the papers in *JME* can be divided pretty easily into three separate categories: chemical synthesis of molecules thought necessary for the origin of life, comparisons of DNA or protein sequences, and abstract mathematical models.

IN THE BEGINNING

The origin-of-life question is tremendously important and interesting. Biology must ultimately deal with the question: even if life evolves by natural selection acting on variation, how did life get there in the first place? Publications concerned with the chemical synthesis of molecules thought to be necessary for the origin of life constitute about 10 percent of all papers in *JME*.

The story of Stanley Miller is one of the best known in all of modern science. As a young graduate student after World War II working in the laboratory of Nobel laureate Harold Urey at the University of Chicago, Miller wanted to determine what chemicals might have been present billions of years ago on the ancient, lifeless earth. He knew that hydrogen is the predominant element in the universe. When hydrogen reacts with carbon, nitrogen, and oxygen—common elements on the earth—it forms methane, ammonia, and water. So Miller decided to see what chemicals could be produced by a simulated atmosphere that contained methane, ammonia, water vapor, and hydrogen.[1]

Methane, ammonia, water vapor, and hydrogen are generally unreactive. Miller knew that, to get the gases to produce potentially interesting chemicals, he would have to pump some energy into the system to jumble things up. One source of energy that would have been available on the old earth is lightning. So Miller constructed an apparatus in his laboratory that contained the gases he expected to be present on the early earth, plus a pool of water, as well as sparking electrodes to simulate lightning.

Miller boiled the water and sparked the mixture of gases for about a week. During that time an oily, insoluble tar built up on the sides of the flask, and the pool of water became more and more reddish as material accumulated in it. At the end of the week Miller analyzed the mixture of chemicals dissolved in the water and saw that it contained several kinds of amino acids. The result electrified the world. Since amino acids are the building blocks of proteins, it appeared at first blush that the materials for making the machines of life would be plentiful on the early earth. Excited scientists had no difficulty imagining that natural processes might induce amino acids to come together to form proteins, that some of the proteins would catalyze important chemical reactions, that the proteins would get trapped inside small cell-like membranes, that nucleic acids would be produced by similar processes, and that gradually the first truly self-replicating cell would be born. As with Mary Shelley's fictional Frankenstein, it appeared that electricity coursing through inanimate matter could indeed produce life.

Other experimenters rushed to build on the seminal work of Stanley Miller. He had detected a few different types of amino acids in his experiment, but living organisms contain twenty different kinds. Other researchers varied Miller's experimental conditions. The mix of gases in the simulated atmosphere was altered, the source of energy was changed from an electric spark to ultraviolet radiation (to simulate sunlight) or very strong pulses of pressure (to simulate explosions.) More sophisticated analytical methods detected chemicals that were present in very small amounts. Sustained effort by a number of workers eventually paid off; almost all of the twenty naturally occurring types of amino acids have been detected in origin-of-life experiments.

Other successes were reported in the early years of research on the origin of life. Perhaps the most notable achievement was by the laboratory of Juan Oró. They showed that the simple chemical hydrogen cyanide would react with itself to yield a number of products including adenine, which is a component of one of the building blocks of nucleic acids. The result cracked open DNA and RNA as targets for chemical investigation of life's origin. Over the years other components of nucleic acids—the other "bases," as well as the sugar ribose which forms part of RNA—were produced by chemical simulation experiments.

In light of these well-publicized successes an outsider can be excused for feeling a sense of shock when he stumbles across pessimistic reviews of origin-of-life research in the professional literature, such as one written by Klaus Dose, a prominent worker in the field. In his assessment of the state of the problem, Dose pulls no punches.

> More than 30 years of experimentation on the origin of life in the fields of chemical and molecular evolution have led to a better perception of the immensity of the problem of the origin of life on Earth rather than to its solution. At present all discussions on principal theories and experiments in the field either end in stalemate or in a confession of ignorance.[2]

What leads a professional in the field to such a bleak view, especially after the progress in the heady days following Miller's trailblazing experiment? It turns out that the successes, although real, paper over a plethora of problems that can only be appreciated when you move beyond the simple chemical production of some of the bare components of life. Let's look at a few of those problems.

Making the molecules of life by chemical processes outside of a cell is actually rather easy. Any competent chemist can buy some chemicals from a supply company, weigh them in the correct proportion, dissolve them in an appropriate solvent, heat them in a flask for a predetermined amount of time, and purify the desired chemical produce away from unwanted chemicals produced by side reactions. Not only can amino acids and nucleotides—the building blocks—be made, but a chemist can then take these and produce the buildings themselves: proteins and nucleic acids. As a matter of fact, the process for doing this has been automated, and machines that mix and react chemicals to give proteins and nucleic acids are sold by a number of commercial firms. Any undergraduate can read the instruction manual and produce a long piece of DNA—perhaps the gene coding for a known protein—in a day or two.

Most readers will quickly see the problem. There were no chemists four billion years ago. Neither were there any chemical supply houses, distillation flasks, nor any of the many other devices that the modern chemist uses daily in his or her laboratory, and which are necessary to get good results. A convincing origin-of-life scenario requires that intelligent direction of the chemical reactions be minimized as far as possible. Nonetheless, the involvement of some intelligence is unavoidable. Reasonable guesses about what substances were available on the early earth—such as Stanley Miller made—are a necessary starting point. The trick for the researcher is to choose a probable starting point, then keep his hands off.

As an analogy, suppose a famous chef said that random natural processes could produce a chocolate cake. In his effort to prove it, we would not begrudge him taking whole plants—including wheat, cacao, and sugar cane—and placing them near a hot spring, in the hope that the heated water would extract the right materials and cook them. But we would become a little wary if the chef bought refined flour, cocoa, and sugar at the store, saying that he didn't have time to wait for the hot water to extract the components from the plants. We would shake our heads if he then switched his experiment from a hot spring to an electric oven, to "speed things up." And we would walk away if he then measured the amounts of the components carefully, mixed them in a bowl, placed them in a pan, and baked them in his oven. The results would have nothing to do with his original idea that natural processes could produce a cake.

The experiment that Stanley Miller reported in 1952 stunned the world. As Miller has readily explained, however, that experiment was not the first such one he tried. Earlier he had set up his apparatus in a somewhat different manner and found that some oil was formed, but no amino acids. Since he thought amino acids would be the most interesting chemicals to find, he jiggled the apparatus around in hopes of producing them. Of course, if conditions on the ancient earth actually resembled Miller's unsuccessful attempts, then in reality no amino acids would have been produced.

Moreover, joining many amino acids together to form a protein with a useful biological activity is a much more difficult chemical problem than forming amino acids in the first place. The major problem in hooking amino acids together is that, chemically, it involves the removal of a molecule of water for each amino acid joined to the growing protein chain. Conversely, the presence of water strongly inhibits amino acids from forming proteins. Because water is so abundant on the earth, and because amino acids dissolve readily in water, origin-of-life researchers have been forced to propose unusual scenarios to get around the water problem. For example, a scientist named Sidney Fox proposed that perhaps some amino acids got washed up from the primordial ocean onto a very hot surface, such as the rim of an active volcano. There, the story goes, they would be heated above the boiling point of water; with the water gone, the amino acids could join together. Unfortunately, other workers had earlier shown that heating dry amino acids gives a smelly, dark brown tar, but no detectable proteins. Fox, however, demonstrated that if an extra-large portion of one of three different amino acids is added to a mix of purified amino acids and heated in a laboratory oven, then the amino acids do join. But even then they do not join to give proteins—the structure they form is chemically different. So Fox and collaborators called the structures "proteinoids," then went on to show that the proteinoids had some interesting properties, including modest catalytic abilities, that were reminiscent of real proteins.

The scientific community has remained deeply skeptical of these experiments. As with our imaginary baker, a heavy odor of investigator involvement hangs over proteinoids. The special circumstance needed to make them—hot, dry conditions (putatively representing rare spots such as volcano rims) with exact amounts of already-purified amino acids weighed out in advance—casts dark shadows over the relevance of the experiments. Worse, because proteinoids are not really proteins, the considerable problem of producing authentic proteins remains. In his book reviewing the difficulties of origin-of-life theories, Robert Shapiro notes that work on proteinoids has produced a startling unanimity of opinion:

> [The proteinoid theory] has attracted a number of vehement critics, ranging from chemist Stanley Miller . . . to Creationist Duane Gish. On perhaps no other point in origin-of-life theory could we find such harmony between evolutionists and Creationists as in opposing the relevance of the experiments of Sidney Fox.[3]

Other researchers have proposed some other ways whereby amino acids might join to give proteins. All suffer more or less from the problems that plague proteinoids, and none has attracted much support from the scientific community.

THE RNA WORLD

In the 1980s a scientist named Thomas Cech showed that some RNA has modest catalytic abilities.[4] Because RNA, unlike proteins, can act as a template and so potentially can catalyze its own replication, it was proposed that RNA—not protein—started earth on the road to life. Since Cech's work was reported, enthusiasts have been visualizing a time when the world was soaked with RNA on its way to life; this model has been dubbed "the RNA world." Unfortunately, the optimism surrounding the RNA world ignores known chemistry. In many ways the RNA-world fad of the 1990s is reminiscent of the Stanley Miller phenomenon during the 1960s: hope struggling valiantly against experimental data.

Imagining a realistic scenario whereby natural processes may have made proteins on a prebiotic earth—although extremely difficult—is a walk in the park compared to imagining the formation of nucleic acids such as RNA. The big problem is that each nucleotide "building block" is itself built up from several components, and the processes that form the components are chemically incompatible. Although a chemist can make nucleotides with ease in a laboratory by synthesizing the components separately, purifying them, and then recombining the components to react with each other, undirected chemical reactions overwhelmingly produce undesired products and shapeless goop on the bottom of the test tube. Gerald Joyce and Leslie Orgel—two scientists who have worked long and hard on the origin of life problem—call RNA "the prebiotic chemist's nightmare." They are brutally frank:

> Scientists interested in the origins of life seem to divide neatly into two classes. The first, usually but not always molecular biologists, believe that RNA must have been the first replicating molecule and that chemists are exaggerating the difficulties of nucleotide synthesis. . . . The second group of scientists are much more pessimistic. They believe that the de novo appearance of oligonucleotides on the primitive earth would have been a near miracle. (The authors subscribe to this latter view). Time will tell which is correct.[5]

Even if the miracle-like coincidence should occur and RNA be produced, however, Joyce and Orgel see nothing but obstacles ahead. In an article section entitled "Another Chicken-and-Egg Paradox" they write the following:

> This discussion . . . has, in a sense, focused on a straw man: the myth of a self-replicating RNA molecule that arose de novo from a soup of random polynucleotides. Not only is such a notion unrealistic in light of our current understanding of prebiotic chemistry, but it should strain the credulity of even an optimist's view of RNA's catalytic potential. . . . Without evolution it appears unlikely that a self-replicating ribozyme could arise, but without some form of self-replication there is no way to conduct an evolutionary search for the first, primitive self-replicating ribozyme.

In other words, the miracle that produced chemically intact RNA would not be enough. Since the vast majority of RNAs do not have useful catalytic properties, a second miraculous coincidence would be needed to get just the right chemically intact RNA.

Origin-of-life chemistry suffers heavily from the problem of road kill, discussed in the last chapter. Just as there is no absolute barrier to a groundhog crossing a thousand-lane highway during rush hour, so there is no absolute barrier to the production of proteins, nucleic acids, or any other biochemical by imaginable, natural chemical processes; however, the slaughter on the highway is unbearable. The solution of some prebiotic chemists is a simple one. They release a thousand groundhogs by the side of the road, and note that one makes it across the first lane. They then put a thousand fresh groundhogs in a helicopter, fly them to the beginning of lane two, and lower them onto the highway. When one survives the crossing from lane two to lane three, they helicopter another thousand to the edge of lane three. Proponents of the RNA world, who start their experiments with long, purified, investigator-synthesized RNA, fly the groundhogs out to lane 700 and watch as one crosses to lane 701. It is a valiant effort, but if they ever reach the other side, the victory will be quite hollow.

Scientists working on the origin of life deserve a lot of credit; they have attacked the problem by experiment and calculation, as science should. And although the experiments have not turned out as many hoped, through their efforts we now have a clear idea of the staggering difficulties that would face an origin of life by natural chemical processes.

In private many scientists admit that science has no explanation for the beginning of life.[7] On the other hand many scientists think that given the origin of life, its subsequent evolution is easy to envision, despite the major difficulties outlined in this book. The reason for this peculiar circumstance is that while chemists try to test origin-of-life scenarios by experiment or calculation, evolutionary biologists make no attempt to test evolutionary scenarios at the molecular level by experiment or calculation. As a result, evolutionary biology is stuck in the same frame of mind that dominated origin-of-life studies in the early fifties, before most experiments had been done: imagination running wild. Biochemistry has, in fact, revealed a molecular world that stoutly resists explanation by the same theory so long applied at the level of the whole organism. Neither of Darwin's starting points—the origin of life, and the origin of vision—has been accounted for by his theory. Darwin never imagined the exquisitely profound complexity that exists even at the most basic levels of life.

Over the years the *Journal of Molecular Evolution* has published origin-of-life research concerning many questions, such as the following: Could other amino acids not found by Miller also be produced? What if carbon dioxide predominated in the ancient atmosphere instead of methane? Could nucleotides other than modern ones have started life? Such questions have been addressed in *JME* in papers with titles like "Prebiotic Syntheses in Atmospheres Containing CH_4, CO, and CO_2,"[8] "Radiolysis of Aqueous Solutions of Hydrogen Cyanide (pH 6): Compounds of Interest in Chemical Evolution Studies,"[9] "Alternative Bases in the RNA World: The Prebiotic Synthesis of Urazole and Its Ribosides,"[10] and "Cyclization of Nucleotide Analogues as an Obstacle to Polymerization."[11] These are interesting questions for scientists, but they do not begin to answer the challenge to evolution posed by blood clotting, cellular transport, or disease fighting.

THE MISSING PAPERS

The second category of papers commonly found in the *Journal of Molecular Evolution*, accounting for about 5 percent of the total, concerns mathematical models for evolution or new mathematical methods for comparing and interpreting sequence data. This includes papers with titles such as "A Derivation of All Linear Invariants for a Nonbalanced Transversion Model"[12] and "Monte Carlo Simulation in Phylogenies:

An Application to Test the Constancy of Evolutionary Rates."[13] Although useful for understanding how gradual processes behave over time, the mathematics *assumes* that real-world evolution is a gradual, random process; it does not (and cannot) demonstrate it.

By far the largest category of papers published in *JME,* accounting for more than 80 percent of all manuscripts, is that of sequence comparisons. A sequence comparison is an amino-acid-by-amino-acid comparison of two different proteins, or a nucleotide-by-nucleotide comparison of two different pieces of DNA, noting the positions at which they are identical or similar, and the places where they are not.

When methods were developed in the 1950s to determine the sequences of proteins, it became possible to compare the sequence of one protein with another. A question that was immediately asked was whether analogous proteins in different species, like human hemoglobin and horse hemoglobin, had the same amino acid sequence. The answer was intriguing: horse and human hemoglobin were very similar, but not identical. Their amino acids were the same in 129 out of 146 positions in one of the protein chains of hemoglobin, but different in the remaining positions. When the sequences of the hemoglobins of monkey, chicken, frog, and others became available, their sequences could be compared with human hemoglobin and with each other. Monkey hemoglobin had 5 differences with that of humans; chickens had 26 differences; and frogs had 46 differences. These similarities were highly suggestive. Many researchers concluded that similar sequences strongly supported descent from a common ancestor.

For the most part it was shown that analogous proteins from species that were already thought to be closely related (like man and chimp, or duck and chicken) were pretty similar in sequence, and proteins from species thought to be distantly related (such as skunk and skunk cabbage) were not that similar. In fact, for some proteins one could correlate the amount of sequence similarity with the estimated time since various species were thought to have last shared a common ancestor, and the correlation was quite good. Emile Zuckerkandl and Linus Pauling then proposed the molecular clock theory, which says that the correlation is caused by proteins accumulating mutations over time. The molecular clock has been vigorously debated since it was proposed, and many issues surrounding it are still contended. Overall, however, it remains a viable possibility.

In the late 1970s, quick and easy methods became available for sequencing DNA. Thus one could study not only the sequence of a protein but also the gene for the protein, as well as the DNA surrounding
the gene that contained control regions and other features. Genes from
higher organisms were shown to contain interruptions (called introns)
in the coding sequence. Some genes had dozens of introns; other
genes just one or two. So now a biochemist could publish comparisons of the sequences of the introns in genes from different species, as
well as studies of the total number of introns, their relative positioning
in the gene, their length and base composition, and a dozen other factors. Other aspects of the genetic apparatus could also be compared:
the position of genes relative to other genes, the frequency with which
one type of nucleotide was found next to another, the number of
chemically modified nucleotides, and so forth. Very many such papers
have been published over the years in the *Journal of Molecular Evolution*, including "Examination of Protein Sequence Homologies: IV.
Twenty-Seven Bacterial Ferredoxins,"[14] "Evolution of α- and ß-Tubulin
Genes as Inferred by the Nucleotide Sequences of Sea Urchin cDNA
Clones,"[15] "Phylogeny of Protozoa Deduced from 5S rRNA Sequences,"[16] and "Tail-to-Tail Orientation of the Atlantic Salmon Alpha-
and Beta-Globin Genes."[17]

Although useful for determining possible lines of descent, which is
an interesting question in its own right, comparing sequences cannot
show how a complex biochemical system achieved its function—the
question that most concerns us in this book.[18] By way of analogy, the
instruction manuals for two different models of computer put out by
the same company might have many identical words, sentences, and
even paragraphs, suggesting a common ancestry (perhaps the same
author wrote both manuals), but comparing the sequences of letters in
the instruction manuals will never tell us if a computer can be produced step-by-step starting from a typewriter.

The three general topics of papers published in *JME*—the origin of
life, mathematical models of evolution, and sequence analyses—have
included many intricate, difficult, and erudite studies. Does such valuable and interesting work contradict this book's message? Not at all.
To say that Darwinian evolution cannot explain everything in nature is
not to say that evolution, random mutation, and natural selection do
not occur; they have been observed (at least in cases of microevolu-

tion) many different times. Like the sequence analysts, I believe the evidence strongly supports common descent. But the root question remains unanswered: What has caused complex systems to form? No one has ever explained in detailed, scientific fashion how mutation and natural selection could build the complex, intricate structures discussed in this book.

In fact, *none* of the papers published in *JME* over the entire course of its life as a journal has ever proposed a detailed model by which a complex biochemical system might have been produced in a gradual, step-by-step Darwinian fashion. Although many scientists ask how sequences can change or how chemicals necessary for life might be produced in the absence of cells, no one has ever asked in the pages of *JME* such questions as the following: How did the photosynthetic reaction center develop? How did intramolecular transport start? How did cholesterol biosynthesis begin? How did retinal become involved in vision? How did phosphoprotein signaling pathways develop? The very fact that none of these problems is even addressed, let alone solved, is a very strong indication that Darwinism is an inadequate framework for understanding the origin of complex biochemical systems.

To take up the questions raised in this book, one would need to find papers with titles such as "Twelve Intermediate Steps Leading to the Bacterial Photosynthetic Reaction Center," "A Proto-Cilium Could Generate a Power Stroke Sufficient to Turn a Cell by Ten Degrees," "Intermediates in Adenosine Biosynthesis Effectively Mimic Adenosine Itself in RNA Function," and "A Primitive Clot Made of Randomly Aligned Fibers Would Block Circulation in Veins Smaller Than 0.3 Millimeters." But the papers are missing. Nothing remotely like this has been published.

Perhaps we can understand why detailed models are missing from *JME* by asking what a real scientific investigation of mousetrap evolution by an enthusiastic young scientist would look like. He would first have to think of a precursor to the modern mousetrap, one that was simpler. Suppose he started with just a wooden platform? No, that won't catch mice. Suppose he started with a modern mousetrap that has a shortened holding bar? No, if the bar is too short it wouldn't reach the catch, and the trap would spring uselessly while he was holding it. Suppose he started with a smaller trap? No, that wouldn't explain the complexity. Suppose the parts developed individually for other functions—such as a Popsicle stick for the platform, a clock spring for the trap spring, and so on—and then accidentally got together? No, their previous functions would leave them unfit for trapping mice, and he'd still have to explain how they gradually developed into a mousetrap. With his tenure evaluation coming up, a smart young scientist would switch his interests to more tractable topics.

Attempts to explain the evolution of highly specified, irreducibly complex systems—either mousetraps or cilia or blood clotting—by a gradualistic route have so far been incoherent, as we have seen in previous chapters. No scientific journal will publish patently incoherent papers, so no studies asking detailed questions of molecular evolution are to be found. Calvin and Hobbes stories can sometimes be spun by ignoring critical details, as Russell Doolittle did when imagining the evolution of blood clotting, but even such superficial attempts are rare. In fact, evolutionary explanations even of systems that do not appear to be irreducibly complex, such as specific metabolic pathways, are missing from the literature. The reason for this appears to be similar to the reason for the failure to explain the origin of life: a choking complexity strangles all such attempts.

SEARCHING HIGH AND LOW

There are scores of journals devoted to biochemical research. Although *JME* carries articles concerning molecular evolution exclusively, other journals carry such articles also, mixed in with research on other topics. Perhaps, then, it is a mistake to conclude too much based just on a survey of *JME*. Perhaps other, nonspecialized journals publish research on the origins of complex biochemical systems. To see if *JME* is simply the wrong place to look, let's take a quick look at a prestigious journal that covers a broad range of biochemical topics: the *Proceedings of the National Academy of Sciences*.

Between 1984 and 1994 *PNAS* published about twenty thousand papers, the large majority of which were in the life sciences. Every year the journal compiles an index in which it lists the year's papers by category. The index shows that in those ten years, about 400 papers were concerned with molecular evolution.[19] This is approximately one-third as many papers as the *Journal of Molecular Evolution* published over the same time period. The number of papers on the topic published yearly by *PNAS* has increased significantly, going from about 15 in 1984 to about 100 in 1994; clearly this is a growth area. But the great majority (about 85 percent) are concerned with sequence analysis, just as most papers in *JME* were, passing over the fundamental question of how. About 10 percent of the molecular evolution papers are mathematical studies—either new methods to improve sequence comparisons or highly abstract models. No papers were published in *PNAS* that proposed detailed routes by which complex biochemical structures might have developed. Surveys of other biochemistry journals show the same result: sequences upon sequences, but no explanations.

Perhaps if there are no answers in journals then we should look in books. Darwin proposed his revolutionary theory in a book, and so did Newton. The advantage of a book is that it gives the author a lot of room to develop his or her ideas. Setting a new idea in context, bringing in appropriate examples, explaining a lot of detailed steps, meeting many anticipated objections—all of this can take a fair amount of space. A good example in the modern evolution literature is a book called *The Neutral Theory of Molecular Evolution* by Motoo Kimura.[20] In the book he had the room to explain his idea that most sequence changes that occur in DNA and proteins do not affect the way they do their jobs; the mutations are neutral. A second example is *The Origins of Order* by Stuart Kauffman, who argues that the origins of life, metabolism, genetic programs, and body plans are all beyond Darwinian explanation but may arise spontaneously through self-organization.[21] Neither book explains biochemical structures: Kimura's work has to do simply with sequences, and Kauffman's is a mathematical analysis. But perhaps in one of the libraries of the world there is a book that tells us how specific biochemical structures came to be.

Unfortunately, a computer search of library catalogs shows there is no such book. That isn't too surprising in this day and age; even books like Kimura's and Kauffman's that propose new theories are usually preceded by papers on the topic that are first published in scientific journals. The absence of papers on the evolution of biochemical structures in the journals just about kills any chance of there being a book published on the matter.

During a computer search for books on biochemical evolution, you come across a number of juicy titles. For example, a book by John

Gillespie was published in 1991 with the enticing name *The Causes of Molecular Evolution*. But it does not concern specific biochemical systems. It is, like Kauffman's, a mathematical analysis that leaves out all of the specific features of organisms, reducing them to mathematical symbols and then manipulating the symbols. Nature is blanched. (I should add that, of course, mathematics is an extremely powerful tool. But math is useful to science only when the assumptions the mathematical analysis starts with are true.)

Another book, published the same year, is *Evolution at the Molecular Level*.[22] Although it sounds promising, it is not a book by someone proposing a new idea. It's one of the many academic books that are collections of articles by different authors, each treating a particular area in not much more depth than a journal article. Inevitably the contents of the book pretty closely resemble the contents of the journals: a lot of sequences, some math, and no answers.

A somewhat different type of book is one that reports the results from a scientific meeting. Cold Spring Harbor Laboratories on Long Island has sponsored a number of meetings on various topics throughout the years. A meeting was held there in 1987 on the topic of "Evolution of Catalytic Function," and about one hundred papers by the participants were published as a book.[23] As is typical of meeting books, about two-thirds of the papers are simply overviews of what was going on in the author's lab at the time, with little or no attempt to tie it into the theme of the book. Of the remaining papers, most are sequence analyses, and some are concerned with prebiotic chemistry or simple catalysts (not the complex machinery of known organisms).

The search can be extended, but the results are the same. There has never been a meeting, or a book, or a paper on details of the evolution of complex biochemical systems.

ACCULTURATION

Many scientists are skeptical that Darwinian mechanisms can explain all of life, but a large number do believe it. Since we have just seen that the professional biochemical literature contains no papers or books that explain in detail how complex systems might have arisen, why is Darwinism nonetheless credible with many biochemists? A large part of the answer is that they have been taught as part of their biochemical training that Darwinism is true. To understand both the success of Darwinism as orthodoxy and its failure as science at the molecular level, we have to examine the textbooks that are used to teach aspiring scientists.

One of the most successful texts of biochemistry over the past several decades was first written in 1970 by Albert Lehninger, a professor of biophysics at Johns Hopkins University, and has been updated several times over the years. On the first page of the first chapter of his first textbook, Lehninger mentions evolution. He asks why the biomolecules that occur in virtually all cells appear to be extraordinarily well fitted to their tasks:

> In this chapter, the first in a series of 12 devoted to the structures and properties of the major classes of biomolecules, we shall develop the idea that biomolecules should be studied from two points of view. We must of course examine their structure and properties as we would those of nonbiological molecules, by the principles and approaches used in classical chemistry. But we must also examine them in the light of the hypothesis that biomolecules are the products of evolutionary selection, that they may be the fittest possible molecules for their biological function.[24]

Lehninger, a fine teacher, was passing on to his students the worldview of biochemical professionals—that evolution is important for understanding biochemistry, that it is one of just two "points of view" by which they must study the molecules of life. Although a callow student might take Lehninger's word for it, a dispassionate observer would look for evidence of evolution's importance to the study of biochemistry. An excellent place to start is the book's index.

Lehninger provided a very detailed index in his book to help students readily find information. Many topics in the index have multiple entries, because they must be considered in various contexts. For example, ribosomes have 21 entries in the index of Lehninger's first edition; photosynthesis has 26 entries; the bacterium E. coli has 42 entries; and under "proteins" are entered 70 references. In all, there are nearly 6,000 entries in the index, but only 2 under the heading of "evolution." The first citation is in a discussion of the sequences of proteins; as discussed earlier, however, although sequence data can be used to infer relationships, they cannot be used to determine how a

complex biochemical structure originated. Lehninger's second reference is to a chapter on the origin of life in which he discusses proteinoids and other topics that have not stood the test of time.

With just 2 citations out of 6,000, Lehninger's teacherly advice to
his students concerning the importance of evolution to their studies is
belied by his index. In it Lehninger included virtually everything of relevance to biochemistry. Apparently, though, evolution is rarely a relevant topic.

Lehninger published a new edition of his text in 1982; its index
contains just 2 references to evolution out of 7,000 entries. After
Lehninger died in 1986, Michael Cox and David Nelson of the University of Wisconsin updated and rewrote the 1982 text. In the preface
the new authors include the following under a list of goals:

> To project a clear and repeated emphasis on major themes, especially
> those relating to evolution, thermodynamics, regulation, and the relation-
> ship between structure and function.[25]

Indeed, in the index of the new edition there are 22 references to evolution out of a total of 8,000, an increase of more than tenfold from
the last edition. But when we get past origin-of-life chemistry and sequence comparisons (the two references in Lehninger's earlier text),
we find that the new edition uses the word *evolution* as a wand to wave
over mysteries. For example, one citation is to "evolution, adaptation
of sperm whale." When we flip to the indicated page, we learn that
sperm whales have several tons of oil in their heads which becomes
more dense at colder temperatures. This allows the whale to match the
density of the water at the great depths where it often dives and so
swim more easily. After describing the whale the textbook remarks,
"Thus we see in the sperm whale a remarkable anatomical and biochemical adaptation, perfected by evolution."[26] But that single line is
all that's said! The whale is stamped "perfected by evolution," and
everybody goes home. The authors make no attempt to explain how
the sperm whale came to have the structure it has.

The extra references to evolution in the newest edition of the
Lehninger text can all be fit into three categories: sequence similarities, comments on the ancestry of cells, and pious but unsupported at-
tributions of a feature to evolution. But none of these, even in principle, can tell us how molecular machinery arose step by step. In no

instance is a detailed route given by which any complex biochemical system might have arisen in a Darwinian manner.

A survey of thirty biochemistry textbooks (summarized in Table 8–1) used in major universities over the past generation shows that many textbooks ignore evolution completely. For example, Thomas Devlin of Jefferson University in Philadelphia wrote a biochemistry

TABLE 8–1

REFERENCE TO EVOLUTION IN THE INDEXES OF BIOCHEMISTRY TEXTBOOKS

Author	Year	Publisher	Total Number of Entries in Index	Entries Referring to Evolution
Lehninger	1970	Worth	6,000	2
Lehninger	1982	Worth	7,000	2
Lehninger et al.	1993	Worth	8,000	22
Devlin	1982	John Wiley & Sons	3,500	0
Devlin	1986	John Wiley & Sons	2,500	0
Devlin	1992	Wiley-Liss	5,000	0
Stryer	1975	Freeman	3,000	0
Stryer	1981	Freeman	4,000	0
Stryer	1988	Freeman	4,000	14
Stryer	1995	Freeman	4,000	9
Voet & Voet	1990	John Wiley & Sons	9,000	12
Voet & Voet	1995	John Wiley & Sons	10,000	13
Mathews & van Holde	1990	Benjamin Cummings	6,000	9
Horton et al.	1993	Prentice Hall	4,500	11
Moran et al.	1994	Prentice Hall	9,000	12
Zubay	1983	Addison Wesley	5,000	1
Zubay	1988	Macmillan	5,000	3
Zuday	1993	Wm. C. Brown	6,000	19
Zubay et al.	1995	Wm. C. Brown	7,000	2
Armstrong & Bennett	1979	Oxford University	2,500	0
Armstrong	1983	Oxford University	3,000	0
Armstrong	1989	Oxford University	4,000	0
Scheve	1984	Allyn and Bacon	3,000	0
Abeles et al.	1992	Jones and Bartlett	4,500	0
Garrett & Grisham	1995	Harcourt Brace	6,000	5
Wood et al.	1981	Benjamin Cummings	4,000	1
Conn & Stumpf	1976	John Wiley & Sons	2,500	0
Conn et al.	1987	John Wiley & Sons	2,500	1
Kuchel & Ralston	1988	McGraw-Hill	3,500	0
Gilbert	1992	McGraw-Hill	1,000	0

textbook that was first published by John Wiley & Sons in 1982; new editions followed in 1986 and 1992. The first edition has about 2,500 entries in its index; the second also has 2,500; and the third has 5,000. Of these, the number referring to evolution are zero, zero, and zero, respectively. A textbook by Frank Armstrong of North Carolina State University, published by Oxford University Press, is the only recent book to include an historical chapter reviewing important developments in biochemistry, beginning with the synthesis of urea by Friedrich Wöhler in 1828. The chapter does not mention Darwin or evolution. In three editions Armstrong's book has found it unnecessary to mention evolution in its index. Another textbook published by John Wiley & Sons has one citation to evolution in its index out of a total of about 2,500. It refers to a sentence on page 4: "Organisms have evolved and adapted to changing conditions on a geological time scale and continue to do so."[27] Nothing else is said.

Some textbooks make a concerted effort to inculcate in students an evolutionary view of the world. For example, a textbook by Voet and Voet contains a marvelous, full-color drawing nicely capturing the orthodox position.[28] The top third of the drawing shows a volcano, lightning, an ocean, and little rays of sunlight, to suggest how life started. The middle of the picture has a stylized drawing of a DNA molecule leading out from the origin of life ocean and into a bacterial cell, to show how life developed. The bottom third of the picture—no kidding—is like the Garden of Eden, depicting a number of animals that have been produced by evolution milling about. Included in the throng are a man and a woman (the woman is offering the man an apple), both especially attractive and in the buff. This undoubtedly adds to the interest for students, but the drawing is a tease. The implicit promise that the secrets of evolution will be uncovered is never consummated.[29]

Many students learn from their textbooks how to view the world through an evolutionary lens. However, they do not learn how Darwinian evolution might have produced any of the remarkably intricate biochemical systems that those texts describe.

HOW DO YOU KNOW?

How do we know what we say we know—not in some deep philosophical sense, but on a practical, everyday level? On any particular day you might tell someone that you know your living room is painted green, that you know the Philadelphia Eagles are going to win the Super Bowl, that you know the earth goes around the sun, that you know democracy is the best form of government, that you know the way to San Jose. Clearly these different assertions are based on different ways of knowing. What are they?

The first way to know something is, of course, through personal experience. You know that your living room is painted green because you've been in your living room and saw that it was green. (I won't worry here about things like how you know you aren't dreaming or insane or such.) Similarly you know what a bird is, how gravity works (again, in an everyday sense), and how to get to the nearest shopping mall, all by direct experience.

The second way to know things is by authority. That is, you rely on some source of information, believing it to be reliable, when you have no experience of your own. So almost every person who has gone to school believes that the earth goes around the sun, even though very few people would be able to tell you how anybody could even detect that motion. You are relying on authority if, when asked if you know the way to San Jose, you answer yes and pull out a map. You might be able to personally test the map's reliability by using it to navigate to San Jose, but until you do you are relying on authority. Many people believe democracy is superior to other forms of government even though they haven't lived under any other type. They rely on the authority of textbooks and politicians, and perhaps on verbal or pictorial descriptions of what it's like in other societies. Of course other societies do the same, and most of their defenders rely on authority.

But how about those Eagles? How do you know they are going to win it all this year? If pressed you might admit that no sports commentator has picked them to win, so you aren't relying on authority. Furthermore, you have no firsthand information that, say, some of the players are training secretly under a Zen master, who promises to greatly increase their agility. You are not basing it on their performance in the recent past, which has been mediocre to abysmal. If really pressed you might point to successes in the distant past (like their championships in 1948, 1949, and 1960, or their Super Bowl appearance in 1981) and say that you just know that they're due for success this year. So in fact you do not know that the Eagles are going to win this year; it was just a figure of speech. Your assertion is based on neither experience nor authority. It is bluster.

Scientists are people, too, so we can ask how scientists know what they say they know. Like everybody else, scientists know things either through their own experience or through authority. In the 1950s, Watson and Crick saw a diffraction pattern produced by shining X-rays on fibers of DNA and, using their mathematical abilities, determined that DNA was a double helix. They knew by doing, from their own experience. As an undergraduate I learned DNA is a double helix, but I have never done an experiment to show it; I rely on authority. All scientists rely on authority for almost all of their scientific knowledge. If you ask a scientist how she knows about the structure of cholesterol, or the behavior of hemoglobin, or the role of vitamins, she will almost always point you to the scientific literature rather than to her own records of what she has done in her laboratory.

The nice thing about science is that authority is easy to locate: it's in the library. Watson and Crick's work on DNA structure can be tracked down and read in *Nature*. The structure of cholesterol and other things can be found there as well. So we can say we know the structure of DNA or cholesterol based on scientific authority if papers on those topics are in the literature. If James Watson or a Presidential Science Commission decreed that DNA was made of green cheese, however, but didn't publish supporting evidence in the literature, then we could not say that a belief in cheesy DNA was based on scientific authority. Scientific authority rests on published work, not on the musings of individuals. Moreover, the published work must also contain pertinent evidence. If Watson published a bare statement about the curdled composition of DNA in a paper largely devoted to something else, but provided no relevant support, then we still have no scientific authority to back up the claim.

Molecular evolution is not based on scientific authority. There is no publication in the scientific literature—in prestigious journals, specialty journals, or books—that describes how molecular evolution of any real, complex, biochemical system either did occur or even might have occurred. There are assertions that such evolution occurred, but absolutely none are supported by pertinent experiments or calculations. Since no one knows molecular evolution by direct experience, and since there is no authority on which to base claims of knowledge, it can truly be said that—like the contention that the Eagles will win the Super Bowl this year—the assertion of Darwinian molecular evolution is merely bluster.

"Publish or perish" is a proverb that academicians take seriously. If you do not publish your work for the rest of the community to evaluate, then you have no business in academia (and if you don't already have tenure, you will be banished). But the saying can be applied to theories as well. If a theory claims to be able to explain some phenomenon but does not generate even an attempt at an explanation, then it should be banished. Despite comparing sequences and mathematical modeling, molecular evolution has never addressed the question of how complex structures came to be. In effect, the theory of Darwinian molecular evolution has not published, and so it should perish.

CHAPTER 9

INTELLIGENT DESIGN

WHAT'S GOING ON?

The impotence of Darwinian theory in accounting for the molecular basis of life is evident not only from the analyses in this book, but also from the complete absence in the professional scientific literature of any detailed models by which complex biochemical systems could have been produced, as shown in Chapter 8. In the face of the enormous complexity that modern biochemistry has uncovered in the cell, the scientific community is paralyzed. No one at Harvard University, no one at the National Institutes of Health, no member of the National Academy of Sciences, no Nobel prize winner—no one at all can give a detailed account of how the cilium, or vision, or blood clotting, or any complex biochemical process might have developed in a Darwinian fashion. But we are here. Plants and animals are here. The complex systems are here. All these things got here somehow: if not in a Darwinian fashion, then how?

Clearly, if something was not put together gradually, then it must have been put together quickly or even suddenly. If adding individual pieces does not continuously improve the function of a system, then multiple pieces have to be added together. Two ways to rapidly assemble complex systems have been proposed by scientists in recent years. Let's briefly consider those proposals, and then look in depth at a third alternative.

The first alternative to gradualism has been championed by Lynn Margulis. In place of a Darwinian view of progress by competition and strife, she proposes advancement by cooperation and symbiosis. Organisms in her view aid one another, join forces, and accomplish together what they could not accomplish separately. While still a graduate student she brought this idea to bear on problems of cell structure. Although initially patronized and ridiculed, Margulis eventually won grudging acceptance—and then acclaim (she was elected to the National Academy of Sciences)—for her idea that parts of the cell were once free-living organisms.

The eukaryotic cell, as we have seen, is chock full of complex molecular machines tidily separated into many discrete compartments. The biggest compartment is the nucleus, which could be seen even with the crude microscopes of the seventeenth century. Smaller compartments were not discovered until improved microscopes became available in the later nineteenth and twentieth centuries. One of the smaller compartments is the mitochondrion.

Perhaps I should say that many of the smaller compartments are mitochondria: the typical cell contains about two thousand of them, and they occupy a total of about 20 percent of the cell's volume. Each of the little compartments contains machinery necessary to capture the energy of foodstuffs and store it in a chemically stable, yet readily available, form. The mitochondrial mechanisms that do this are quite complex. The system uses a flow of acid to power its machines, which shuttles electrons among a half-dozen carriers, requiring an exquisitely delicate interaction between many components.

Mitochondria are roughly the same size and shape as some free-living bacterial cells. Lynn Margulis proposed that at one time on the ancient earth a larger cell "swallowed" a bacterial cell, but did not digest it. Rather, the two cells—one now living inside the other—adapted to the situation. The smaller cell received nutrients from the larger one and, in return, passed on some of the stored chemical energy it made to the larger cell. When the larger cell reproduced, the smaller one did too, and its descendants continued to reside inside the host. Over time the symbiotic cell lost many of the systems that free-living cells need, and specialized more and more in providing energy for its host. Eventually it became a mitochondrion.

The stifled laughs and smirks that greeted Margulis's proposal slowly faded when new sequencing techniques, developed after she proposed the theory, showed that mitochondrial proteins more closely resemble bacterial proteins than host cell proteins. Other resemblances between mitochondria and bacteria were then noticed. Furthermore, proponents of the symbiotic origin of mitochondria pointed to symbiotic cells in contemporary organisms to support their theory. For example, a species of flatworm has no mouth because it never has to eat—it contains photosynthetic algae that supply its energy! Such pieces of evidence have carried the day. Margulis's theory concerning mitochondria has now become textbook orthodoxy.

Periodically over the last two decades Margulis and other scientists have proposed that other cellular compartments are the result of symbiosis. These proposals are not so widely accepted. For purposes of argument, however, let's suppose that the symbiosis Margulis envisions was in fact a common occurrence throughout the history of life. The important question for us biochemists is, can symbiosis explain the origin of complex biochemical systems?

Clearly it cannot. The essence of symbiosis is the joining of two separate cells, or two separate systems, *both of which are already functioning*. In the mitochondrion scenario, one preexisting viable cell entered a symbiotic relationship with another such cell. Neither Margulis nor anyone else has offered a detailed explanation of how the preexisting cells originated. Proponents of the symbiotic theory of mitochondria explicitly assume that the invading cells could already produce energy from foodstuffs; they explicitly assume that the host cell already was able to maintain a stable internal environment that would benefit the symbiont.

Because symbiosis starts with complex, already-functioning systems, it cannot account for the fundamental biochemical systems we have discussed in this book. Symbiosis theory may have important points to make about the development of life on earth, but it cannot explain the ultimate origins of complex systems.

The second alternative to Darwinian gradualism proposed in recent years is known as "complexity theory" and has been championed by Stuart Kauffman. In brief, complexity theory states that systems with a large number of interacting components spontaneously organize themselves into ordered patterns. Sometimes there are several patterns available to the complex system, and "perturbations" of the system can cause it to switch from one pattern to the other. Kauffman proposes that chemicals in the prebiotic soup organized themselves into complex metabolic pathways. He further proposes that the switch between different cell "types" (like when a developing organism starts with just a fertilized egg but then goes on to make liver cells, skin cells, etc.) is a perturbation of a complex system and results from the self-organization he envisions.

The above explanation may sound a bit fuzzy. Some of the fuzz is no doubt due to my modest powers of description. But a good deal is due to the fact that complexity theory began as a mathematical concept to describe the behavior of some computer programs, and its proponents have not yet succeeded in connecting it to real life. Rather, the chief mode of argumentation so far has been for proponents to point to the behavior of a computer program and assert that the computer behavior resembles the behavior of a biological system. For example, Kauffman writes about changes (which he calls mutations) in some computer programs he has written:

> Most mutations have small consequences because of the system's [change-resisting] nature. A few mutations, however, cause larger cascades of change. Poised systems will therefore typically adapt to a changing environment gradually, but if necessary, they can occasionally change rapidly. These properties are observed in organisms.[1]

In other words, some small changes in a computer program cause large changes in the program's output (typically a pattern of dots on a computer screen), so perhaps small changes in DNA can produce large, coordinated biological changes. The argument never goes further than that. No proponent of complexity theory has yet gone into a laboratory, mixed a large variety of chemicals in a test tube, and looked to see if self-sustaining metabolic pathways spontaneously organize themselves. If they ever do try such an experiment, they will merely be repeating the frustrating work of origin-of-life scientists who have gone before them—and who have seen that complex mixtures yield a lot of muck on the sides of a flask, and not much else.

In his book on the subject Kauffman muses that complexity theory might explain not only the origin of life and metabolism, but also body shapes, ecological relationships, psychology, cultural patterns, and economics.[2] The vagueness of complexity, though, has started to turn off early boosters of the theory. *Scientific American* ran favorable articles over a number of years (one authored by Kauffman himself). On its cover, however, the June 1995 issue asked, "Is Complexity a Sham?" Inside was an article entitled "From Complexity to Perplexity" that noted the following:

> Artificial life, a major subfield of complexity studies, is "fact-free science," according to one critic. But it excels at generating computer graphics.

Indeed, some proponents see great significance in the fact that they can write short computer programs which display images on the screen that resemble biological objects such as a clam shell. The implication is that it doesn't take much to make a clam. But a biologist or biochemist would want to know, if you opened the computer clam, would you see a pearl inside? If you enlarged the image sufficiently, would you see cilia and ribosomes and mitochondria and intracellular transport systems and all the other systems that real, live organisms need? To ask the question is to answer it. In the article, Kauffman observes that "At some point artificial life drifts off into someplace where I cannot tell where the boundary is between talking about the world— I mean, everything out there—and really neat computer games and art forms and toys." More people are beginning to think that the drifting point occurs very early.

For the sake of argument, however, let us suppose that complexity theory is true—that complex mixtures somehow organized themselves, and that had something to do with the origin of life. Granted its premises, can complexity theory explain the complex biochemical systems we have discussed in this book? I don't believe so. The complex, interacting mixture of chemicals it envisions might have occurred before life developed (again, though, there is virtually no evidence to support even this), but it would not have mattered once cellular life began. The essence of cellular life is regulation: The cell controls how much and what kinds of chemicals it makes; when it loses control, it dies. A controlled cellular environment does not permit the serendipitous interactions between chemicals (always unspecified) that Kauffman needs. Because a viable cell keeps its chemicals on a short leash, it would tend to *prevent* new, complex metabolic pathways from organizing by chance.

Let's further suppose that the pattern of genes that are turned on and off in a cell, corresponding to different cell types, can switch according to the theories of Stuart Kauffman. (Different cell types form when different genes are turned on or off. For example, the genes for hemoglobin—the protein that carries oxygen to tissues—are turned on in cells that make red blood cells, but are turned off in other cells.) Although there is no evidence for it, let us say that complexity theory has something to do with the switch that turns one cell into a red blood cell and another into a nerve cell. Can this explain the origin of complex biochemical systems? No. Like symbiosis theory, this aspect of complexity theory requires preexisting, already functional systems. So if a cell turns off almost all genes except the ones to make hemoglobin, it might turn into a red blood cell; if another cell turns on another set of genes, it might make the proteins characteristic of a nerve cell. But no eukaryotic cell can turn on preexisting genes and suddenly make a bacterial flagellum, because no preexisting proteins in the cell interact in that way. The only way a cell could make a flagellum is if the structure were already coded for in its DNA. In fact, Kauffman never claims that such new and complex structures can be produced suddenly according to complexity theory.

Complexity theory may yet make important contributions to mathematics, and it may still make modest contributions to biochemistry. But it cannot explain the origin of the complex biochemical structures that undergird life. It doesn't even try.

DETECTION OF DESIGN

Imagine a room in which a body lies crushed, flat as a pancake. A dozen detectives crawl around, examining the floor with magnifying glasses for any clue to the identity of the perpetrator. In the middle of the room, next to the body, stands a large, gray elephant. The detectives carefully avoid bumping into the pachyderm's legs as they crawl, and never even glance at it. Over time the detectives get frustrated with their lack of progress but resolutely press on, looking even more closely at the floor. You see, textbooks say detectives must "get their man," so they never consider elephants.

There is an elephant in the roomful of scientists who are trying to explain the development of life. The elephant is labeled "intelligent design." To a person who does not feel obliged to restrict his search to unintelligent causes, the straightforward conclusion is that many biochemical systems were designed. They were designed not by the laws of nature, not by chance and necessity; rather, they were *planned*. The designer knew what the systems would look like when they were completed, then took steps to bring the systems about. Life on earth at its most fundamental level, in its most critical components, is the product of intelligent activity.

The conclusion of intelligent design flows naturally from the data itself—not from sacred books or sectarian beliefs. Inferring that biochemical systems were designed by an intelligent agent is a humdrum process that requires no new principles of logic or science. It comes simply from the hard work that biochemistry has done over the past forty years, combined with consideration of the way in which we reach conclusions of design every day. Nonetheless, saying that biochemical systems were designed will certainly strike many people as strange, so let me try to make it sound less strange.

What is "design"? Design is simply the *purposeful arrangement of parts*. With such a broad definition we can see that anything *might* have been designed. Suppose that as you drive to work one bright morning, you observe a burning car by the side of the road—its front end pushed in, broken glass all around. About twenty feet from the car you see a motionless body lying in a heap. Stamping on the brakes, you pull over to the side of the road. You rush up to the body, grab a wrist to feel for a pulse, and then notice that a young man with a minicam is standing behind a nearby tree. You yell to him to call an ambulance, but he keeps on filming. Turning back to the body, you notice that it is smiling at you. The uninjured actor explains that he is a graduate student in the department of social work and is doing research on the willingness of motorists to come to the aid of injured strangers. You glare at the grinning charlatan as he stands and wipes the fake blood off his face. You then help him to achieve a more realistic look and walk away contentedly as the cameraman runs off to call an ambulance.

The apparent accident was designed; a number of parts were purposely arranged to look like a mishap. Other, less noticeable events could be designed also: The coats on a rack in a restaurant may have been arranged by the owner before you came in. The trash and tin cans along the edge of a highway may have been placed by an artist trying to make some obscure environmental statement. Apparently chance meetings between people might be the result of a grand design (conspiracy theorists thrive on postulating such designs). On the campus of my university there are sculptures that, if I saw them lying beside the road, I would guess were the result of chance blows to a piece of scrap metal, but they were designed.

The upshot of this conclusion—that anything could have been purposely arranged—is that we cannot know that something has *not* been designed. The scientific problem then becomes, how do we confidently detect design? When is it reasonable to conclude, in the absence of firsthand knowledge or eyewitness accounts, that something has been designed? For discrete physical systems—if there is not a gradual route to their production—design is evident when a number of separate, interacting components are ordered in such a way as to accomplish a function beyond the individual components.[3] The greater the specificity of the interacting components required to produce the function, the greater is our confidence in the conclusion of design.

This can be seen clearly in examples from diverse systems. Suppose that you and your spouse are hosting another couple one Sunday afternoon for a game of Scrabble. When the game ends, you leave the room for a break. You come back to find the Scrabble letters lying in the box, some face up and some face down. You think nothing of it until you notice that the letters facing up read, "TAKE US OUT TO DINNER CHEAPSKATES." In this instance you immediately infer design, not even bothering to consider that the wind or an earthquake or your pet cat might have fortuitously turned over the right letters. You infer design because a number of separate components (the letters) are ordered to accomplish a purpose (the message) that none of the components could do by itself. Furthermore, the message is highly specific; changing several of the letters would make it unreadable. For the same reason, there is no gradual route to the message: one letter does not give you part of the message, a few more letters does not give a little more of the message, and so on.

Despite my inability to recognize design in the sculptures around campus, it is often easy to recognize design in other pieces of artwork here. For example, the gardeners arrange flowers near the student center to spell out the name of the university. Even if you had not seen them working, you could easily tell that the flowers had been purposely arranged. For that matter, if you came across flowers deep in the woods that clearly spelled out the name "LEHIGH," you would have no doubt that the pattern was the result of intelligent design.

Design can most easily be inferred for mechanical objects. While walking through a junkyard you might observe separated bolts and screws and bits of plastic and glass—most scattered, some piled on top of each other, some wedged together. Suppose your eye alighted on a pile that seemed particularly compact, and when you picked up a bar sticking out of the pile, the whole pile came along with it. When you pushed on the bar it slid smoothly to one side of the pile and pulled an attached chain along with it. The chain in turn yanked a gear which turned three other gears which turned a rod, spinning it smoothly. You quickly conclude that the pile was not a chance accumulation of junk but was designed (that is, was put together in that order by an intelligent agent), because you see that the components of the system interact with great specificity to do something.

Systems made entirely from natural components can also evince design. For example, suppose you are walking with a friend in the woods. All of a sudden your friend is pulled high in the air and left dangling by his foot from a vine attached to a tree branch. After cutting him down you reconstruct the trap. You see that the vine was wrapped around the tree branch, and the end pulled tightly down to the ground. It was securely anchored to the ground by a forked branch. The branch was attached to another vine—hidden by leaves—so that, when the trigger-vine was disturbed, it would pull down the forked stick, releasing the spring-vine. The end of the vine formed a loop with a slipknot to grab an appendage and snap it up into the air. Even though the trap was made completely of natural materials you would quickly conclude that it was the product of intelligent design.

For a simple artificial object such as a steel rod, the context is often important in concluding design. If you saw the rod outside a steel plant, you would infer design. Suppose however, that you traveled in a rocket ship to a barren alien planet that had never been explored. If you saw dozens of cylindrical steel rods lying on the side of a volcano, you would need more information before you could be sure that alien geological processes—natural for that planet—had not produced the rods. In contrast, if you found dozens of mousetraps near the volcano, you would apprehensively look over your shoulder for signs of the designer.

In order to reach a conclusion of design for something that is not an artificial object (for example, an arrangement of vines and sticks in the woods to make a trap), or to reach a conclusion of design for a system composed of a number of artificial objects, *there must be an identifiable function of the system.* One has to be careful, though, in defining the function. A sophisticated computer can be used as a paper weight; is that its function? A complex automobile can be used to help dam a stream; is that what we should consider? No. In considering design, the function of the system we must look at is the one that requires *the greatest amount of the system's internal complexity.* We can then judge how well the parts fit the function.[4]

The function of a system is determined from the system's internal logic: the function is not necessarily the same thing as the purpose to which the designer wished to apply the system. A person who sees a mousetrap for the first time might not know that the manufacturer expected it to be used for catching mice. He might use it instead for a defense against burglars or as a warning system for earthquakes (if the vibrations would set off the trap), but he still knows from observing how the parts interact that it was designed. Similarly, someone might try to use a lawnmower as a fan or as an outboard motor. But the function of the equipment—to rotate a blade—is best defined by its internal logic.

WHO'S THERE?

Inferences to design do not require that we have a candidate for the role of designer. We can determine that a system was designed by examining the system itself, and we can hold the conviction of design much more strongly than a conviction about the identity of the designer. In several of the examples above, the identity of the designer is not obvious. We have no idea who made the contraption in the junkyard, or the vine trap, or why. Nonetheless, we know that all of these things were designed because of the ordering of independent components to achieve some end.

The inference to design can be made with a high degree of confidence even when the designer is very remote. Archeologists digging for a lost city might come across square stones, buried dozens of feet under the earth, with pictures of camels and cats, griffins and dragons. Even if that were all they found, they would conclude that the stones had been designed. But we can go even further than that. I was a teenager when I saw *2001: A Space Odyssey*. To tell the truth I really didn't care for the movie; I just didn't get it. It started out with monkeys beating each other with sticks, then shifted to a space flight with a homicidal computer, and ended up with an old man spilling a drink and an unborn child floating in space. I'm sure it had some profound meaning, but we scientific types don't catch on quickly to artsy stuff.

There was one scene, however, that I did get quite easily. The first space flight had landed on the moon, and an astronaut was going out to explore. In his meanderings he came across a smoothly shaped obelisk that towered against the moonscape. I, the astronaut, and the rest of the audience immediately understood, with no words necessary, that the object was designed—that some intelligent agent had been to the moon and formed the obelisk. Later the movie showed us that there were aliens on the planet Jupiter, but we couldn't tell that from the obelisk. For all we knew by looking at the object itself, it might have been designed by space aliens, angels, humans from the past (whether Russians or inhabitants of the lost civilization of Atlantis) who could fly through space, or even by one of the other astronauts on the flight (who, as a practical joke, might have stowed it away and put it on the moon ahead of the astronaut who later discovered it). If the plot had actually developed along any of these lines, the audience would not be able to say the plot was contradicted by the appearance of the obelisk. If the movie had contrived to assert that the obelisk was not designed, however, the audience would have hooted till the projectionist turned the film off.

The conclusion that something was designed can be made quite independently of knowledge of the designer. As a matter of procedure, the design must first be apprehended before there can be any further question about the designer. The inference to design can be held with all the firmness that is possible in this world, without knowing anything about the designer.

ON THE EDGE

Anyone can tell that Mt. Rushmore was designed—but, as the king of Siam often said, this too shall pass. As time marches and rains fall and winds gust, Mt. Rushmore will change its shape. Millennia in the future, people may pass the mountain and see just the barest hint of faces in the rocks. Could a person conclude that an eroded Mt. Rushmore had been designed? It depends. The inference to design requires the identification of separate components that have been ordered to accomplish a purpose, and the strength of the inference is not an easy matter to quantify. An eroded Mt. Rushmore might give future archeologists fits if they could only see what looked like an ear, a nose, a bottom lip, and maybe a chin, each from a different presidential image. The parts really aren't ordered to each other and might be simply an unusual rock formation.

There appears to be the face of a man on the surface of the moon. One can point to darkened areas that look like eyes and a mouth. This might have been designed, perhaps by aliens, but the number and specificity of the components is not sufficient to determine if the purpose that is ascribed to the pattern was actually intended. Italy may have been intentionally designed to look like a boot, but maybe not. There is not enough data to reach a confident conclusion. The *National Enquirer* once ran a story purporting to show a human face on the surface of Mars; however, the resemblance was only slight. In such cases we can just say that, like anything, it could have been designed, but we cannot tell for sure.

As the number and quality of the components that fit together to form the system increases, we can be more and more confident of the conclusion of design. A few years ago it was reported that an image of Elvis was formed by mold growing on the refrigerator of a lady from Tennessee. Again, the resemblance could be seen, but it was slight. Suppose, however, that the resemblance was actually very good. Suppose that the image was made up not only of black mold. Suppose that there was also *Serratia marcescens*—a bacterium that grows in red sheets. And suppose there were colonies of the yeast *Saccharomyces cerevisiae*, which are shiny white. And there was also *Pseuodomonas aeruginosa*, which is green, and *Chromobacterium violaceum*, which is purple, and *Staphylococcus aureus*, which is yellow. And suppose the green microorganisms were growing in the shape of Elvis's pants, and the purple bacteria formed his shirt. And very small dots of alternating red and white bacteria gave his face a flesh-colored look.

In fact, suppose the bacteria and mold on the refrigerator formed an image of Elvis that was well nigh identical to one of those velvet posters of him that you see in variety stores. Can we then conclude that the image was designed? Yes we can—with the same confidence that we conclude the dimestore posters were designed.

If the "man in the moon" had a beard and ears and eyeglasses and eyebrows we would conclude that it was designed. If Italy had button-holes and shoelaces and if Sicily closely resembled a soccer ball, with colored stripes and a logo, we would think that they were designed. As the number or quality of the parts of an interacting system increase, our judgment of design increases also and can reach certitude. It is hard to quantify these things.[5] But it is easy to conclude that a system of such detail as the completed bacterial Elvis was designed.

BIOCHEMICAL DESIGN

It is easy to see design in Elvis posters, mousetraps, and Scrabble mes-sages. But biochemical systems aren't inanimate objects; they're part of living organisms. Can living biochemical systems be intelligently de-signed? It wasn't too long ago that life was thought to be made of a special substance, different from the stuff that comprised nonliving objects. Friedrich Wöhler debunked that idea. For a long while after-ward, the complexity of life defeated most attempts to understand and manipulate it. In recent decades, however, biochemistry has made such great strides that basic changes in living organisms are being de-signed by scientists. Let's take a look at a few examples of biochemical design.

When the blood-clotting system misfires, a wayward clot can block blood flow through the heart, endangering life. In current treatment a naturally occurring protein is injected into the patient to help break up the clot. But the natural protein has some drawbacks, so innovative re-searchers are trying to *make a new protein* in the laboratory that can do a better job.[6] Briefly, the strategy is the following (Figure 9–1). Many of the proteins of the blood-clotting system are activated by other factors that clip a piece of the target protein, activating it. The piece that is clipped, however, is targeted by just its activator and no other. Plasminogen—the precursor of plasmin, the protein that breaks up blood clots—contains a target that is clipped only very slowly, after the clot has formed and healing begins. To treat a heart attack, though, plasmin is needed immediately at the site of the blood clot that is inhibiting circulation.

In order to make plasmin available immediately at the right place, the gene for plasminogen has been isolated by researchers and altered. The part of the gene coding for the site in plasminogen that is cleaved to activate the protein is replaced. It is replaced by the part of a gene

FIGURE 9–1

(1) THE GENE FOR PLASMINOGEN IS ISOLATED. (IN THE FIGURE THE AMINO ACIDS, NOT THE DNA, THAT THE GENE CODES FOR ARE SHOWN.) (2) THE SECTION OF THE GENE THAT CODES FOR THE AREA OF THE PROTEIN THAT IS CUT SLOWLY DURING ACTIVATION IS TAKEN OUT. (3) THE SECTION OF ANOTHER GENE THAT CODES FOR A PROTEIN REGION THAT IS CUT RAPIDLY BY THROMBIN IS PUT INTO THE PLASMINOGEN GENE. (4) A DESIGNED, HYBRID GENE NOW EXISTS THAT WILL, WHEN PLACED IN A CELL, PRODUCE A PLASMINOGEN THAT IS RAPIDLY ACTIVATED BY THROMBIN.

1)   —DCGKPQVEPKKC**PGR**VVGGCVAHPHSWPWQ—

2)   —DCGKPQVEPKKC-   -VGGCVAHPHSWPWQ—

-TTKIKPRI-

↓

3)   —DCGKPQVEPKKC-   -VGGCVAHPHSWPWQ—

4)   —DCGKPQVEPKKC**TTKIKPRI**VGGCVAHPHSWPWQ—

for another component of the blood-clotting pathway (such as plasma thromboplastin antecedent, or PTA) that is cleaved rapidly by thrombin. Now the idea is this: the engineered plasminogen, carrying the thrombin-cleavable piece, will quickly be cut and activated in the close vicinity of a clot, because thrombin is present at the clot site. But the activity that is quickly released is not that of PTA; rather, it is plasmin. If such a protein were quickly injected into a heart attack victim, the hope is that the plasmin would help him or her recover with minimum permanent damage.

The new protein is the product of intelligent design. Someone with knowledge of the blood-clotting system sat down at his desk and sketched out a route to produce a protein that would combine the clot-dissolving properties of plasmin with the rapid-activation property of proteins that are cleaved by thrombin. The designer knew what the end product of his work was going to do, and he worked to achieve that goal. After the plan was drawn up, the designer (or his graduate student) went into the laboratory and took steps to carry out the plan. The result is a protein that no one in the world has ever seen before— a protein that will carry out the plan of the designer. Biochemical systems can indeed be designed.

Intelligent design of biochemical systems is really quite commonplace these days. In order to supply diabetics with hard-to-get human insulin, researchers a decade ago isolated the human insulin gene. They placed it into a piece of DNA that could survive in a bacterial cell and grew up the modified bacteria. The bacteria's cellular machinery then produced human insulin, which was isolated and used to treat patients. Some laboratories are now modifying higher organisms by incorporating altered DNA directly into their cells. Designed plants that resist frost or insect pests have been around for a while now; somewhat newer is the engineering of cows that give milk containing large amounts of useful proteins. (The people who do this by injecting extraneous genes into cow embryos like to call themselves "pharmers," short for pharmaceutical farmers.)

It might be observed that although the systems described above are examples of biochemical design, in each case the designer did no more than rearrange pieces of nature; he or she did not produce a new system from scratch. That is true, but it probably won't be true for very long. Scientists today are actively working on uncovering the secrets of what gives proteins their special activity. Progress has been slow but steady. It won't be long before proteins are made from scratch, designed for specific, novel purposes. Even more impressively, new chemical systems are being developed by organic chemists to mimic the activities of life. This has been played up in the popular media as "synthetic life." Although that is a gross exaggeration designed to sell magazines, the work does show that an intelligent agent can design a system exhibiting biochemical-like properties without using the biochemicals known to occur in living systems.

In recent years some scientists have even begun to design new biochemicals using the principles of microevolution—mutation and selection.[7] The idea is simple: chemically make a large number of different pieces of DNA or RNA, then pull out of the mix the few pieces that have a property that the designer wants, such as the ability to bind to a vitamin or protein. This is done by mixing solid particles to which the vitamin or protein has been attached with a solution containing the mix of DNA or RNA pieces, and then washing away the solution. Pieces of DNA or RNA that bind the vitamin or protein remain stuck to the solid; all the pieces which don't bind are washed away. After selecting the right pieces the experimenter uses enzymes to make many copies of them. Gerald Joyce, a leader in the field, likens the process to selective breeding: "If one wants a redder rose or a fluffier Persian, one chooses as breeding stock those individuals that best exemplify the desired trait. So, too, if one wants a molecule that exhibits a particlar chemical trait, then one selects from a large population of molecules those individuals that best manifest the property."[8] Like selective breeding, the method has the advantages of microevolution, but also has its limitations. Simple biochemical activities can be produced, but not the complicated systems we have discussed in this book.

In many ways this technique is like the clonal selection of antibodies, discussed in Chapter 7. Indeed, other scientists are taking advantage of the ability of the immune system to generate antibodies against almost any molecule. The scientists inject an animal with a molecule of interest (for example, a drug) and isolate the antibodies that are made against it. The antibodies can then be used as clinical or commercial reagents to detect the molecule. In some cases antibodies can be produced which behave like simple enzymes[9] (they are called "abzymes"). Both of these approaches—DNA/RNA or antibodies—

promise to find a host of industrial and medical applications in the coming years.

The fact that biochemical systems can be designed by intelligent agents for their own purposes is conceded by all scientists, even Richard Dawkins. In his newest book Dawkins envisions a hypothetical scenario where a leading scientist is kidnaped and forced to work on biological weapons for an evil, militaristic country.[10] The scientist gets help by encoding a message in the DNA sequence of an influenza virus: he infects himself with the altered virus, sneezes on a crowd of people, and patiently waits for the flu to spread around the world, confident that other scientists will isolate the virus, sequence its DNA, and decipher his code. Since Dawkins agrees that biochemical systems can be designed, and that people who did not see or hear about the designing can nonetheless detect it, then the question of whether a given biochemical system was designed boils down simply to adducing evidence to support design.

We must also consider the role of the laws of nature. The laws of nature can organize matter—for example, water flow can build up silt sufficiently to dam a portion of a river, forcing it to change course. The most relevant laws are those of biological reproduction, mutation, and natural selection. If a biological structure can be explained in terms of those natural laws, then we cannot conclude that it was designed. Throughout this book, however, I have shown why many biochemical systems cannot be built up by natural selection working on mutations: no direct, gradual route exists to these irreducibly complex systems, and the laws of chemistry work strongly against the undirected development of the biochemical systems that make molecules such as AMP. Alternatives to gradualism that work through unintelligent causes, such as symbiosis and complexity theory, cannot (and do not even try to) explain the fundamental biochemical machines of life. If natural laws peculiar to life cannot explain a biological system, then the criteria for concluding design become the same as for inanimate systems. There is no magic point of irreducible complexity at which Darwinism is logically impossible. But the hurdles for gradualism become higher and higher as structures are more complex, more interdependent.

Might there be an as-yet-undiscovered natural process that would explain biochemical complexity? No one would be foolish enough to categorically deny the possibility. Nonetheless, we can say that if there is such a process, no one has a clue how it would work. Further, it would go against all human experience, like postulating that a natural process might explain computers. Concluding that no such process exists is as scientifically sound as concluding that mental telepathy is not possible, or that the Loch Ness monster doesn't exist. In the face of the massive evidence we do have for biochemical design, ignoring that evidence in the name of a phantom process would be to play the role of the detectives who ignore an elephant.

With these preliminary questions cleared out of the way, we can conclude that the biochemical systems discussed in Chapters 3 through 6 were designed by an intelligent agent. We can be as confident of our conclusion for these cases as we are of the conclusion that a mousetrap was designed, or that Mt. Rushmore or an Elvis poster were designed. There is no question of degree for those systems, such as for the man in the moon or the shape of Italy. Our ability to be confident of the design of the cilium or intracellular transport rests on the same principles as our ability to be confident of the design of anything: the ordering of separate components to achieve an identifiable function that depends sharply on the components.

The function of the cilium is to be a motorized paddle. In order to achieve this function microtubules, nexin linkers, and motor proteins all have to be ordered in a precise fashion. They have to recognize each other intimately, and interact exactly. The function is not present if any of the components is missing. Furthermore, many more factors besides those listed are required to make the system useful for a living cell: the cilium has to be positioned in the right place, oriented correctly, and turned on or off according to the needs of the cell.

The function of the blood-clotting system is as a strong, but transient barrier. The components of the system are ordered to that end. Fibrinogen, plasminogen, thrombin, protein C, Christmas factor, and the other components of the pathway together do something that none of the components can do alone. When vitamin K is unavailable or antihemophilic factor is missing, the system crashes just as surely as a Rube Goldberg machine fails if a component is missing. The components cut each other in precise places, align with each other in exact ways. They act to form an elegant structure that accomplishes a specific task.

The function of the intracellular transport systems is to carry cargo from one place to another. To do this packages must be labeled, destinations recognized, and vehicles equipped. Mechanisms must be in place to leave one enclosed area of the cell and enter a different enclosed area. The failure of the system leaves a deficit of critical supplies here, a choking surplus there. Enzymes that are useful in a confined area wreak havoc in another area.

The functions of the other biochemical systems I have discussed are readily identifiable, and their interacting parts can be enumerated. Because the functions depend critically on the intricate interactions of the parts we must conclude that they, like a mousetrap, were designed.

The designing that is currently going on in biochemistry laboratories throughout the world—the activity that is required to plan a new plasminogen that can be cleaved by thrombin, or a cow that gives growth hormone in its milk, or a bacteria that secretes human insulin—is analogous to the designing that preceded the blood-clotting system. The laboratory work of graduate students piecing together bits of genes in a deliberate effort to make something new is analogous to the work that was done to cause the first cilium.

MAKING DISTINCTIONS

Just because we can infer that some biochemical systems were designed does not mean that all subcellular systems were explicitly designed. Further, some systems may have been designed, but proving their design may be difficult. The face of Elvis might be clear and distinct while his (assumed) guitar is an impressionistic blur. Detecting design in the cilium might be a piece of cake, but design in another system might be borderline or undetectable. It turns out that the cell contains systems that span the range from obviously designed to no apparent design. Keeping in mind that anything might have been designed, let's take a brief look at a couple of systems where design is hard to see.

The basis of life is the cell, in which the biochemical processes that undergird the cell's existence are cordoned off from the rest of the environment. The structure that encapsulates the cell is called the cell membrane. It is made up mostly of molecules that are chemically similar to the detergents with which we wash our dishes and clothes. The exact type of detergent-like molecules that are used in membranes varies widely from one kind of cell to another: some are longer, some are shorter; some are looser, some are stiffer; some have positive charges, some have negative charges, and some are neutral. Most cells contain a mixture of different types of molecules in their membranes, and the blend can be different for different types of cells.

When detergent molecules find themselves in water, they tend to associate with each other. A good example of this association is seen in the bubbles that slosh around in the washing machine while you're doing laundry. The bubbles consist of very thin layers of detergent (plus some water) in which the molecules are packed side by side. The spherical shape of the bubbles is due to a physical force called surface tension, which acts to reduce the area of the bubble to the smallest area able to accommodate the detergent. If you take the molecules from a cell membrane, purify them away from all the other components of a cell, and dissolve them in water, they will often pack together into a spherical, enclosed shape.

Because these molecules form bubbles on their own, because the association of molecules is indiscriminate, and because a particular individual molecule is not necessary to form a membrane, it is difficult to infer intelligent design from cell membranes. Like the stones in a stone wall, each of the components is easily replaced by a different component. Like the mold on *my* refrigerator, design is not detectable.

Or consider hemoglobin—the protein in our red blood cells that carries oxygen from the lungs to the peripheral tissue. Hemoglobin is made up of four individual proteins stuck together, and each of the four proteins can bind oxygen. Two of the four proteins are identical to each other, as are the other two to each other. It turns out that, because of the way the four component proteins of hemoglobin stick to each other, the first oxygen that hops on binds less strongly than the last three. The difference in the strength of binding oxygen results in a behavior called "cooperativity." Simply put, this means that the amount of oxygen bound by a large number of hemoglobins (as occurs in the blood) does not increase directly with the amount of oxygen in the air. Rather, when the amount of oxygen in the surroundings is low, practically no oxygen binds to hemoglobin—much less than would bind if there were no cooperativity. On the other hand, when the oxygen in the surroundings increases, the amount of oxygen bound to he-moglobin in the blood increases at a very fast rate. This can be thought of as something like a domino effect; it takes some effort to knock over the first domino (bind the first oxygen), but the other dominos then fall down automatically. Cooperativity has important physiological consequences: it allows hemoglobin to become fully saturated where there is a lot of oxygen (such as in the lungs) and to easily dump off the oxygen where it is needed (such as peripheral tissues).

There is also another protein, called myoglobin, that is very similar to hemoglobin except that it has only one protein chain, not four, and therefore binds only one oxygen. The binding of oxygen to myoglobin is not cooperative. The question is, if we assume that we already have an oxygen-binding protein like myoglobin, can we infer intelligent de-sign from the function of hemoglobin? The case for design is weak. The starting point, myoglobin, already can bind oxygen. The behavior of hemoglobin can be achieved by a rather simple modification of the behavior of myoglobin, and the individual proteins of hemoglobin strongly resemble myoglobin. So although hemoglobin can be thought of as a system with interacting parts, the interaction does nothing much that is clearly beyond the individual components of the system. Given the starting point of myoglobin, I would say that hemoglobin shows the same evidence for design as does the man in the moon: in-triguing, but far from convincing.

The final biochemical system is one I already talked about in Chap-ter 7—the system that makes AMP. Concluding design here is like con-cluding that a painting attributed to a famous-but-dead artist is actu-ally a forgery by another person from the same era. Perhaps you see that the painting has the famous artist's name printed on the lower left corner, but the brush strokes, the color combination, the subject mat-ter, the canvas material, and the paint itself are all different.

Because so many successive steps are needed to make AMP, because the intermediates are not used, and because our best chemical knowl-edge argues strongly against the undirected production of the path-way, the case for the design of the AMP pathway appears to be very strong. In theory the conclusion for design here is vulnerable to a Stu-art Kauffman–type scenario; however, complexity theory is currently not much more than a phantom, and the known chemical behavior of molecules strongly dictates against the scenario. Furthermore, the conclusion of intelligent design for other biochemical systems bolsters the credibility of invoking design for this system as well.

Since anything could have been designed, and since we need to adduce evidence to show design, it is not surprising that we can be more successful in demonstrating design with one biochemical system and less successful with another. Some features of the cell appear to be the result of simple natural processes, others probably so. Still other features were almost certainly designed. And with some features, we can be as confident that they were designed as that anything was.

CHAPTER 10

QUESTIONS ABOUT DESIGN

SIMPLE IDEAS

A simple idea can take a surprising length of time to be properly developed, even though the idea is very powerful. Perhaps the most famous example of this is the invention of the wheel. Before the wheel people slogged around in horse-drawn carts that slid on poles, scraping across the ground and generating a lot of friction. Any schoolboy of our time could have advised them to build wagons with wheels, because the schoolboy has learned about wheels. The idea of a wheel is both extremely powerful and, looking back, stunningly simple, and it leads to all sorts of practical advantages in life. Yet the idea was formed and developed only with difficulty.

Another powerful idea is the phonetic alphabet. Phonetic alphabets are comprised of symbols that stand for sounds; by putting together several of these symbols, one gets a symbol string that stands for the sound of a real word. Phonetic alphabets contrast with hieroglyphic writing systems, in which pictorial characters stand for words. In many ways hieroglyphics are a much more natural way to write, especially for someone who is just beginning. Someone who has no knowledge of written communication is much more likely to draw a picture of a dog eating a bone than to write marks on paper in the form of "DOG EAT BONE" and then tell all his friends that the mark resembling a half circle with a line down one side (D) stands for the *sound* "duh," the circle (O) stands for the sound "ahh," and so on. If it were already in place, the more natural hieroglyphic system would tend to prevent a phonetic alphabet from being adopted, even though a phonetic system is actually simpler and much more versatile as language becomes more complex.

In grammar school we learn that in the number 561 the digit 1 stands for 1, but the digit 6 stands for 60, and the digit 5 stands for 500. Because of this little place-value trick, working with numbers becomes so simple that a child can do it. Any ten-year-old who has been properly instructed can add 561 to 427 to get 988, and any twelve-year-old can multiply 41 by 17 to get 697. But try to add or multiply those numbers using Roman numerals! Try to add XXIV to LXXVI to get C (without first converting the Roman numerals to Arabic numerals). Roman numerals were used in Europe until the Middle Ages; consequently, the vast majority of the populace could not do the simple calculations that a modern teller or cashier can do. Simple sums required the talents of specially trained people who earned their living by counting.

SLOUCHING TOWARD DESIGN

The idea of intelligent design is also a simple, powerful, obvious idea that has been sidetracked by competition from, and contamination with, extraneous ideas. From the beginning the chief competitor to a rigorous design hypothesis has been the fuzzy feeling that if something fit our idea of the way things ought to be, then that was evidence of design. The early Greek philosopher Diogenes saw design in the regularity of the seasons:

> Such a distribution would not have been possible without Intelligence, that all things should have their measure: winter and summer and night and day and rain and winds and periods of fine weather; other things also, if one will study them closely, will be found to have the best possible arrangement.[1]

Socrates is said to have observed:

> Is it not to be admired . . . that the mouth through which the food is con-
> veyed should be placed so near the nose and eyes as to prevent the pas-
> sage unnoticed of whatever is unfit for nourishment? And cans't thou still
> doubt, Aristodemus, whether a disposition of parts like this should be the
> work of chance, or of wisdom and contrivance.[2]

Such sentiments, although humanly understandable, are based simply
on the feeling that the world is a jolly place, and not much else. It is
not hard to imagine that if Diogenes lived in Hawaii, where winter
weather does not come, he might easily think that the lack of seasons
was "the best possible arrangement." If Socrates's mouth was next to
his hand we could imagine him saying that was convenient for trans-
ferring food to the mouth. Arguments to design based on the bare assertion of their "rightness" evaporate like the morning dew when faced
with the least skepticism.

Over the course of human history, most learned folks (and even
more unlearned folks) have thought that design was evident in nature.
Up until the time of Darwin, in fact, the argument that the world was
designed was commonplace in both philosophy and science. But the
intellectual soundness of the argument was poor, probably due to lack
of competition from other ideas. The pre-Darwinian strength of the design argument reached its zenith in the writings of the nineteenth-century Anglican clergyman William Paley. An enthusiastic servant of his
God, Paley brought a wide scientific scholarship to bear in his writings
but, ironically, set himself up for refutation by overreaching.

The famous opening paragraph of Paley's Natural Theology shows
the power of the argument and also contains some of the flaws that led
to its later rejection:

> In crossing a heath, suppose I pitched my foot against a *stone*, and were
> asked how the stone came to be there, I might possibly answer, that for
> any thing I knew to the contrary it had lain there for ever; nor would it,
> perhaps, be very easy to show the absurdity of this answer. But suppose I
> had found a *watch* upon the ground, and it should be inquired how the
> watch happened to be in that place, I should hardly think of the answer
> which I had before given, that for any thing I knew the watch might have
> always been there. Yet why should this answer not serve for the watch as

well as for the stone; why is it not as admissible in the second case as in the first? For this reason, and for no other, namely, that when we come to inspect the watch, we perceive—what we could not discover in the stone—that its several parts are framed and put together for a purpose, e.g. that they are so formed and adjusted as to produce motion, and that motion so regulated as to point out the hour of the day; that if the different parts had been differently shaped from what they are, or placed after any other manner or in any other order than that in which they are placed, either no motion at all would have been carried on in the machine, or none which would have answered the use that is now served by it. To reckon up a few of the plainest of these parts and of their offices, all tending to one result: We see a cylindrical box containing a coiled elastic spring, which, by its endeavor to relax itself, turns round the box. We next observe a flexible chain. . . . We then find a series of wheels. . . . We take notice that the wheels are made of brass, in order to keep them from rust; . . . that over the face of the watch there is placed a glass, a material employed in no other part of the work, but in the room of which, if there had been any other than a transparent substance, the hour could not be seen without opening the case. This mechanism being observed—it requires indeed an examination of the instrument, and perhaps some previous knowledge of the subject, to perceive and understand it; but being once, as we have said, observed and understood, the inference we think is inevitable, that the watch must have had a maker—that there must have existed, at some time and at some place or other, an artificer or artificers who formed it for the purpose which we find it actually to answer, who comprehended its construction and designed its use.[3]

Compared with that of the Greeks, Paley's argument is much improved. Although in *Natural Theology* he gives many poor examples of design (akin to Diogenes and Socrates), he also frequently hits the nail on the head. Among other things, Paley writes about discrete systems, such as muscles, bones, and mammary glands, that he believes would cease to function if one of several components were missing. This is the essence of the design argument. However, it must be emphasized for the modern reader that, even at his best, Paley was talking about biological black boxes: systems larger than a cell. Paley's example of a watch, in contrast, is excellent because the watch was not a black box; its components and their roles were known.

SIDETRACKED

Paley expresses the design argument so well that he even earns the respect of dedicated evolutionists. Richard Dawkins's *The Blind Watchmaker* takes its title from Paley's watch analogy but claims that evolution, rather than an intelligent agent, plays the role of the watchmaker:

> Paley drives his point home with beautiful and reverent descriptions of the dissected machinery of life, beginning with the human eye. . . . Paley's argument is made with passionate sincerity and is informed by the best biological scholarship of his day, but it is wrong, gloriously and utterly wrong. . . . If [natural selection] can be said to play the role of watchmaker in nature, it is the *blind* watchmaker. . . . But one thing I shall not do is belittle the wonder of the living "watches" that so inspired Paley. On the contrary, I shall try to illustrate my feeling that here Paley could have gone even further.[4]

Dawkins's feeling toward Paley is that of a conqueror toward a worthy but defeated enemy. Magnanimous in victory, the Oxford scientist can afford to pay tribute to the cleric who shared Dawkins's own concern for complexity in nature. Certainly Dawkins is justified in considering Paley to be defeated; very few philosophers or scientists refer to him anymore. Those that do, like Dawkins, do so only to dismiss rather than engage his argument. Paley has been lumped in with earth-centered astronomy and the phlogiston theory of burning—another loser in science's struggle to explain the world.

But exactly where, we may ask, was Paley refuted? Who has answered his argument? How was the watch produced without an intelligent designer? It is surprising but true that the main argument of the discredited Paley has actually never been refuted. Neither Darwin nor Dawkins, neither science nor philosophy, has explained how an irreducibly complex system such as a watch might be produced without a designer. Instead Paley's argument has been sidetracked by attacks on its injudicious examples and off-the-point theological discussions. Paley, of course, is to blame for not framing his argument more tightly. But many of Paley's detractors are also to blame for refusing to engage his main point, playing dumb in order to reach a more palatable conclusion.

A MIXED BAG

In *Natural Theology* Paley points to biological examples that, he argues, are systems of interacting components like a watch and therefore indicate a designer. Paley's examples are a mixed bag, ranging from the truly astonishing to the mildly interesting to the rather silly, from mechanical systems to instincts to mere shapes. Almost none of his examples has been specifically refuted by demonstrating that the features could arise without a designer, but because for many examples Paley appeals to no principle that would prevent incremental development, people have assumed since Darwin that such gradual development is possible.

Paley is at his best when writing about mechanical systems. Concerning the heart, he observes as follows:

> It is evident that it must require the interposition of *valves*—that the success indeed of its action must depend upon these; for when any one of its cavities contracts, the necessary tendency of the force will be to drive the enclosed blood not only into the mouth of the artery where it ought to go, but also back again into the mouth of the vein from which it flowed. . . . The heart, constituted as it is, can no more work without valves than a pump can.[5]

Here he identifies a function of the system and tells the reader why the heart requires several parts—not just a pump, but also valves.

Paley is mediocre, though, when describing instincts:

> What should induce the female bird to prepare a nest before she lays her eggs? . . . The fulness or distension which she might feel in a particular part of the body, from the growth and solidity of the egg within her, could not possibly inform her that she was about to produce something which, when produced, was to be preserved and taken care of. . . . How should birds know that their eggs contain their young?[6]

The example may be interesting, but it is hard to put your finger on an exact function in this example. Also, many of the component parts of the system (perhaps residing in the bird's brain) are unknown, and so it is a black box.

Paley was probably overtired when writing about fetal development:

The eye is of no use at the time when it is formed. It is an optical instrument made in a dungeon; constructed for the refraction of light to a focus, and perfect for its purpose before a ray of light has had access to it. . . . It is *providing* for the *future*.[7]

In this example Paley invites us to admire simply the timing of an event, not any feature of a particular, identified system.

Paley seems actively to invite ridicule when writing of what he calls compensation:

The short unbending neck of the *elephant* is compensated by the length and flexibility of his proboscis. . . .

The *crane* kind are to live and seek their food among the waters; yet having no web-foot, are incapable of swimming. To make up for this deficiency, they are furnished with long legs for wading, or long bills for groping, or usually with both. This is *compensation*.[8]

Reasoning such as this can provide a rich source of comedy material (he's tall to compensate for being so ugly; she's rich to compensate for being so dumb; and so on), but it does precious little to demonstrate design. To be charitable, Paley may have thought that his strong examples made design inevitable, and he used the weak examples as icing on the cake. He likely didn't anticipate that later opponents would refute his argument by attacking the icing.

REFUTING PALEY

Despite many of his misguided examples, Paley's famous first paragraph concerning the watch is exactly correct—no one would deny that if you found a watch you would immediately, and with certainty, conclude that it had been designed. The reason for the conclusion is just as Paley implied: the ordering of separate components to accomplish a function beyond that of the individual components. The function of the watch is to act as a timekeeping device. Its components are the various gears, springs, chains, and the like that Paley lists.

So far, so good. But if Paley knows what to look for in his mechanical paradigm, why did he go downhill so quickly? Because he got carried away and started to look at the wrong features of the watch.

The problems start when Paley digresses from systems of necessarily interacting components to talk about arrangements that simply fit his idea of the way things ought to be. The first hint of trouble comes in Paley's opening paragraph, when he mentions that the watch's wheels are made of brass to prevent rust. The problem is that the exact material, brass, is not *required* for the watch to function. It might help, but a watch can function with wheels made of almost any hard material— probably even wood or bone. Things only get worse when Paley mentions the glass cover of the watch. Not only is the exact material not required, but the whole component is dispensable: a cover is not necessary for the function of the watch. A watch cover is simply a convenience that has been attached to an irreducibly complex system, not part of the system itself.

Throughout his book Paley strays from the feature of the watch—a system of *interacting* components—that caused him to select it as an example in the first place. As is often the case for the rest of us, too, his argument would have been greatly improved if he had said less.

Because of his indiscretion, Paley's argument over the years has been turned into a straw man to knock down. Instead of dealing with the real complexity of a system (such as a retina or a watch), some defenders of Darwinism are satisfied with offering a story to account for peripheral features. As an analogy, a Darwinian "explanation" for a watch with a cover would start by assuming that a factory already was making a watch without a cover! And then the explanation would go on to show what an improvement a cover would be.

Poor Paley. His modern opponents feel justified in assuming enormously complex starting points (such as a watch or a retina) if they think they can then explain a simple improvement (such as a watch cover or curvature of the eye). No further arguments are made; no explanation is given for the real complexity, the irreducible complexity. The refutation of Paley's overreaching is asserted to be a refutation of Paley's main point, even by those who know better.

THE ARGUMENT AGAINST DESIGN

Just as the argument for intelligent design has been around a long time, arguments against design have, too. The best arguments are made by Darwin and his successors, but some arguments are older than Darwin's theory. The philosopher David Hume argued against design in *Dialogues Concerning Natural Religion,* published in 1779. In *The Blind Watchmaker* Richard Dawkins recalls a dinner conversation with "a well-known atheist" that touched on the subject:

> I said that I could not imagine being an atheist at any time before 1859, when Darwin's *Origin of Species* was published. 'What about Hume?', replied the philosopher. "How did Hume explain the organized complexity of the living world?" I asked. "He didn't," said the philosopher. "Why does it need any special explanation?"[9]

Dawkins goes on to explain:

> As for David Hume himself, it is sometimes said that that great Scottish philosopher disposed of the Argument from Design a century before Darwin. But what Hume did was criticize the logic of using apparent design in nature as *positive* evidence for the existence of a God. He did not offer an alternative explanation for complex biological design.[10]

A modern philosopher, Elliott Sober of the University of Wisconsin, in his book *Philosophy of Biology,* explains Hume's reasoning for us in more detail:

> Hume believes . . . we must ask how similar watches and organisms really are. A moment's reflection shows that they are very dissimilar. Watches are made of glass and metal; they do not breathe, excrete, metabolize, or reproduce. . . . The immediate consequence, of course, is that the design argument is a very weak analogy argument. It is preposterous to infer that organisms have a given property simply because watches happen to have it.[11]

But Sober does not agree with Hume:

> Although Hume's criticism is devastating if the design argument is an argument from analogy, I see no reason why the design argument must be construed in this way. Paley's argument about organisms stands on its own, regardless of whether watches and organisms happen to be similar. The point of talking about watches is to help the reader see that the argument about organisms is compelling.[12]

In other words, David Hume thought that the design argument depended on a close similarity *in accidental details* of biological organisms to other designed objects. But this line of thinking would destroy all analogies, since any two nonidentical objects will differ in more ways than they are similar. For example, by Hume's thinking you could not liken a car to an airplane, even though both are transportation devices, because an airplane has wings and a car does not, and so forth. Sober rejects Hume's thinking because he says the intelligent-design argument is really something called an inference to the best explanation. This simply means that, given a choice between the competing explanations of intelligent design versus unguided natural forces, Paley's argument would seem more likely (at least, says Sober, before Darwin came along).

Sober's conclusion is fine as far as it goes, but he could also have noted that the argument from analogy is still valid; it was just twisted out of shape by Hume. Analogies always are set up so that they either explicitly or (more frequently) implicitly propose that A is like B in a restricted subset of properties. Rust is like tooth decay in that they both start from small spots and work outward, even though tooth decay takes place in living materials, is caused by bacteria, can be inhibited by fluoride, and so on. A Rube Goldberg machine is like the blood-clotting system in that they are both irreducibly complex, even though they have many differences. In order to reach a conclusion based on an analogy, it is only necessary that the deduction flow out of the shared properties: The irreducibly complex Rube Goldberg machine required an intelligent designer to produce it; therefore the irreducibly complex blood-clotting system required a designer also.

Incidentally, even by Hume's criteria, the analogy between a watch and a living organism could be made very strong. Modern biochemistry probably *could* make a watch, or a time-keeping device, out of biological materials—if not now, then certainly in the near future. Many biochemical systems keep time, including the cells that pace the heart, the system that initiates puberty, and the proteins that tell the cell when to divide. Moreover, biochemical components are known that can act as gears and flexible chains, and feedback mechanisms (which are necessary to regulate a watch) are common in biochemistry. Hume's criticism of the design argument that asserts a fundamental difference between mechanical systems and living systems is out of date, destroyed by the advance of science which has discovered the machinery of life.

Sober continues his analysis of Hume:

I now turn to Hume's second criticism of the design argument, which is no more successful than the first. . . . [Hume] contends that if we are to have good reason to think that the organisms in *our* world are the products of intelligent design, then we must have looked at lots of *other* worlds and observed intelligent designers producing organisms there.[13]

Hume is criticizing design as an inductive argument. An example of an induction is the argument that because no pigs have ever been observed to fly, pigs in all probability cannot fly. A conclusion of design based on induction would require that we have experience of living things being designed. Hume thinks that since we have not observed such designing in our world, we must look to other worlds for such an experience. Since we have no knowledge of other worlds, however, then we have no experience to make an induction. Sober believes that Hume's argument is invalid because, again, Sober thinks that intelligent design is actually an inference to the best explanation, not an inductive argument.

And again Sober is right as far as he goes, but he could have gone further. Although Hume's objection to the inductive argument might have been valid in his day, it has been destroyed by the advance of science. Modern biochemistry routinely designs biochemical systems, which are now known to be the basis of life. Therefore we *do* have experience in observing the intelligent design of components of life. There have probably been tens of thousands of experiments in which new biochemical systems were put together, and in the future there will be many, many more.

The failure of David Hume's arguments has required modern opponents of design to advance other rationales for their views. In the remainder of the chapter I will consider the most frequently heard of the modern arguments against design.

THE OUIJA ANALOGY

The philosopher friend of Richard Dawkins who thought that David Hume refuted the argument from design was mistaken in his philosophy as well as in his science. Elliott Sober is more successful with his philosophy, but apparently he is unaware of relevant developments in science. Although he thinks Hume was incorrect, Sober is unsympathetic to claims of intelligent design because he thinks that Darwinian evolution provides a mechanism for the production of life. He does not base his conclusion on published models for the gradual production of irreducibly complex biochemical systems; he does not even consider the molecular basis of life. Rather, he rejects design and embraces Darwinism based primarily (and ironically) on an analogy. He explains in *Philosophy of Biology*:

> The fact that the mutation-selection process has two parts . . . is brought out vividly by Richard Dawkins in his book *The Blind Watchmaker*. Imagine a device that is something like a combination lock. It is composed of a series of disks placed side by side. On the edge of each disk, the twenty-six letters of the alphabet appear. The disks can be spun separately so that different sequences of letters may appear in the viewing window.
>
> How many different combinations of letters may appear in the window? There are 26 possibilities on each disk and 19 disks in all. So there are 26^{19} different possible sequences. One of these is METHINKSITIS-AWEASEL. . . . The probability that METHINKSITISAWEASEL will appear after all the disks are spun is $1/26^{19}$, which is a very small number indeed. . . .
>
> But now imagine that a disk is frozen if it happens to put a letter in the viewing window that matches the one in the target message. The remaining disks that do not match the target then are spun at random, and the process is repeated. What is the chance now that the disks will display the message METHINKSITISAWEASEL after, say, fifty repetitions?
>
> The answer is that the message can be expected to appear after a surprisingly small number of generations of the process. . . .
>
> Variation is generated at random, but selection among variants is nonrandom.[14]

This analogy is intended to illuminate how complex biological systems might have been produced. So we are asked to conclude, based on the spinning-disk analogy, that the cilium evolved step-by-step, that the initial steps in vision could be produced gradually, and so forth. The analogy is offered in lieu of actual evidence that these or other complex systems could have evolved in a Darwinian fashion. And Sober thinks the analogy is so compelling that, based on it, Darwinian evolution now wins as the inference to the best explanation. Dawkins' analogy (which is slightly different in details in his book versus Sober's rendition), though transparently false, appears to have captured the imagination of some philosophers of biology. Besides Sober, Michael Ruse has used a similar example in his book *Darwinism Defended,* as has Daniel Dennett in *Darwin's Dangerous Idea.*

What is wrong with the Dawkins-Sober analogy? Only everything. It purports to be an analogy for natural selection, which requires a function to select. But what function is there in a lock combination that is wrong? Suppose that after spinning the disks for a while, we had half of the letters right, something like MDTUIFKQINIOAFERSCL (every other letter is correct). The analogy asserts that this is an *improvement* over a random string of letters, and that it would somehow help us open the combination lock. But if your life depended on opening a lock that had the combination METHINKSITISAWEASEL, and you tried MDTUIFKQIN-IOAFERSCL, you would be pushing up daisies. If your reproductive success depended on opening the lock, you would leave no offspring. Ironically for Sober and Dawkins, a lock combination is a highly specified, irreducibly complex system that beautifully illustrates why, for such systems, function cannot be approached gradually.

Evolution, we are told by proponents of the theory, is not goal directed. But then, if we start from a random string of letters, why do we end up with METHINKSITISAWEASEL instead of MYDARLINGCLEMENTINE or MEBETARZANYOUBEJANE? As a disk turns, who is deciding which letters to freeze and why? Instead of an analogy for natural selection acting on random mutation, the Dawkins-Sober scenario is actually an example of the very opposite: an intelligent agent directing the construction of an irreducibly complex system. The agent (Sober here) has the target phrase (lock combination) in his mind and guides the result in that direction as surely as a fortune-teller guides a Ouija board. This hardly seems like a secure foundation upon which to build a philosophy of biology.

The fatal problems with the analogy are not difficult to see. It was amusingly skewered by Robert Shapiro, a professor of chemistry at New York University, in his book *Origins: A Skeptic's Guide to the Origin of Life,* which was published seven years before Sober's book.[15] The fact that a distinguished philosopher overlooks simple logical problems that are easily seen by a chemist suggests that a sabbatical visit to a biochemistry laboratory might be in order.

A HOLE IN THE EYE

In discussions about intelligent design, no objection is more frequently repeated than the argument from imperfection. It can be briefly summarized: If there exists an intelligent agent who designed life on earth, then it would have been capable of making life that contained no apparent flaws; furthermore, it would have done so. The argument seems to have a measure of popular appeal. However, it is just the flip side of Diogenes's argument: because something does *not* fit our idea of the way things ought to be, then that is evidence *against* design.

The argument has been echoed by prominent scientists and philosophers, but it is particularly well presented by Kenneth Miller, a professor of biology at Brown University:

> Another way to respond to the theory of intelligent design is to carefully examine complex biological systems for errors that no intelligent designer would have committed. Because intelligent design works from a clean sheet of paper, it should produce organisms that have been optimally designed for the tasks they perform. Conversely, because evolution is confined to modifying existing structures, it should not necessarily produce perfection. Which is it?
>
> The eye, that supposed paragon of intelligent design, offers an answer. We have already sung the virtues of this extraordinary organ, but we have not considered specific aspects of its design, such as the neural wiring of its light-sensing units. These photoreceptor cells, located in the retina, pass impulses to a series of interconnecting cells that eventually pass information to the cells of the optic nerve, which leads to the brain.
>
> An intelligent designer, working with the components of this wiring, would choose the orientation that produces the highest degree of visual quality. No one, for example, would suggest that the neural connections should be placed in front of the photoreceptor cells—thus blocking the light from reaching them—rather than behind the retina.
>
> Incredibly, this is exactly how the human retina is constructed. . . .
>
> A more serious flaw occurs because the neural wiring must poke directly through the wall of the retina to carry the nerve impulses produced by photoreceptor cells to the brain. The result is a blind spot in the retina—a region where thousands of impulse-carrying cells have pushed the sensory cells aside. . . .

> None of this should be taken to suggest that the eye functions poorly. It is a superb visual instrument that serves us exceedingly well. . . . The key to the intelligent design theory . . . is not whether an organ or system works well but whether its basic structural plan is the obvious product of design. The structural plan of the eye is not.[16]

Miller elegantly expresses a basic confusion; the key to intelligent-design theory is *not* whether a "basic structural plan is the obvious product of design." The conclusion of intelligent design for physically interacting systems rests on the observation of highly specified, irreducible complexity—the ordering of separate, well-fitted components to achieve a function that is beyond any of the components themselves. Although I emphasize that one has to examine molecular systems for evidence of design, let's use Miller's essay as a springboard to examine other problems with the argument from imperfection.

The most basic problem is that the argument demands perfection at all. Clearly, designers who have the ability to make better designs do not necessarily do so. For example, in manufacturing, "built-in obsolescence" is not uncommon—a product is intentionally made so it will not last as long as it might, for reasons that supersede the simple goal of engineering excellence. Another example is a personal one: I do not give my children the best, fanciest toys because I don't want to spoil them, and because I want them to learn the value of a dollar. The argument from imperfection overlooks the possibility that the designer might have multiple motives, with engineering excellence oftentimes relegated to a secondary role. Most people throughout history have thought that life was designed despite sickness, death, and other obvious imperfections.

Another problem with the argument from imperfection is that it critically depends on a psychoanalysis of the unidentified designer. Yet the reasons that a designer would or would not do anything are virtually impossible to know unless the designer tells you specifically what those reasons are. One only has to go into a modern art gallery to come across designed objects for which the purposes are completely obscure (to me at least). Features that strike us as odd in a design might have been placed there by the designer for a reason—for artistic reasons, for variety, to show off, for some as-yet-undetected practical purpose, or for some unguessable reason—or they might not. Odd they may be, but they may still be designed by an intelligence. The point of scientific interest is not the internal mental state of the designer but whether one can detect design. In discussing why aliens on other planets might build artificial structures that we could observe from earth, the physicist Freeman Dyson wrote:

> I do not need to discuss questions of motivation, who would want to do these things or why. Why does the human species explode hydrogen bombs or send rockets to the moon? It is difficult to say exactly why.[17]

When considering whether aliens would try to seed other planets with life, Francis Crick and Leslie Orgel wrote:

> The psychology of extraterrestrial societies is no better understood than terrestrial psychology. It is entirely possible that extraterrestrial societies might infect other planets for quite different reasons than those we have suggested.[18]

In their writings, these authors correctly concluded that design could be detected in the absence of information about the designer's motives.

The next problem is that proponents of the argument from imperfection frequently use their psychological evaluation of the designer as positive evidence for undirected evolution. The reasoning can be written as a syllogism:

1. A designer would have made the vertebrate eye without a blind spot.
2. The vertebrate eye has a blind spot.
3. Therefore Darwinian evolution produced the eye.

It is for reasoning such as this that the phrase *non sequitur* was invented. The scientific literature contains no evidence that natural selection working on mutation can produce either an eye with a blind spot, an eye without a blind spot, an eyelid, a lens, a retina, rhodopsin, or retinal. The debater has reached his conclusion in favor of Darwinism based solely on an emotional feeling of the way things ought to be. A more objective observer would conclude only that the vertebrate eye was not designed by a person who is impressed with the argument from imperfection; extrapolation to other intelligent agents is not possible.

Ken Miller's article was not written for *Reader's Digest,* but for *Technology Review.* The readership is technically sophisticated, able to handle abstract scientific concepts, and used to following difficult arguments to solid conclusions. The fact that he offers these readers an argument based on psychology and emotion, instead of hard science, gives the opposite message than he intends about the relative strengths of intelligent design versus evolution.

WHAT DOES IT DO?

There is a subcategory of the no-designer-would-have-done-it-this-way argument that requires a different response. Instead of saying that a useful structure contains flaws that should not have been allowed, the writer points to some feature that has no apparent use at all. Often the feature resembles something that is actually used in other species, and so appears to be something that was in fact used at one time but then lost its function. Vestigial organs play a prominent role in this argument. For example, evolutionary biologist Douglas Futuyma cites the "rudimentary eyes of cave animals; the tiny, useless legs of many snakelike lizards; [and] the vestiges of the pelvis in pythons" as evidence that evolution has occurred.[19] Since I'm a biochemist, I prefer molecular versions of this argument. Ken Miller talks about the several genes that produce different forms of hemoglobin in humans:

> Are the five genes of this complex the elegant products of design, or a series of mistakes of which evolution took advantage? The cluster itself, or more specifically a sixth ß-globin gene in the cluster, provides the answer. This gene is . . . nearly identical to that of the other five genes. Oddly, however, this gene . . . plays no role in producing hemoglobin. Biologists call such regions "pseudogenes," reflecting the fact that however much they may resemble working genes, in fact they are not.[20]

Miller tells the readers that the pseudogene lacks the proper signals to inform the rest of the cell's machinery to make a protein from it. He then concludes as follows:

> The theory of intelligent design cannot explain the presence of nonfunctional pseudogenes unless it is willing to allow that the designer made serious errors, wasting millions of bases of DNA on a blueprint full of junk

and scribbles. Evolution, in contrast, can easily explain them as nothing more than failed experiments in a random process of gene duplication that persist in the genome as evolutionary remnants.[21]

This argument is unconvincing for three reasons. First, because we have not yet discovered a use for a structure does not mean that no use exists. The tonsils were once considered to be useless organs, but an important function in immunity has been discovered for them. A python pelvis might be doing something useful of which we are ignorant. This point also applies on the molecular scale; hemoglobin pseudogenes and other pseudogenes, although they are not used to make proteins, may be used for other things that we don't know about. A couple of potential uses that spring to mind as I sit here at my desk include bonding to active hemoglobin genes during DNA replication in order to stabilize the DNA; guiding DNA recombination events; and aligning protein factors relative to the active genes. Whether any of these are actual duties of the pseudogene for hemoglobin does not matter. The point here is that Miller's assertion rests on assumptions only.

The second reason why Miller's argument fails to persuade is that even if pseudogenes have no function, evolution has "explained" nothing about how pseudogenes arose. In order to make even a pseudocopy of a gene, a dozen sophisticated proteins are required: to pry apart the two DNA strands, to align the copying machinery at the right place, to stitch the nucleotides together into a string, to insert the pseudocopy back into the DNA, and much more. In his article Miller has not told us how any of these functions might have arisen in a Darwinian step-by-step process, nor has he pointed to articles in the scientific literature where we can find the information. He can't do that, because the information is nowhere to be found.

Folks such as Douglas Futuyma, who cite vestigial organs as evidence of evolution, have the same problem. Futuyma never explains how a real pelvis or eye developed in the first place, so as to be able to give rise to a vestigial organ later on, yet both the functioning organ and the vestigial organ require explanation. I do not purport to understand everything about design or evolution—far from it; I just cannot ignore the evidence for design. If I insert a letter into a photocopier, for instance, and it makes a dozen good copies and one copy that has a couple of large smears on it, I would be wrong to use the smeared copy as evidence that the photocopier arose by chance.

Arguments based on perceived faults or vestigial genes and organs run the danger of the argument of Diogenes that the progression of seasons shows intelligent design. It is scientifically unsound to make any assumptions of the way things ought to be.

LONG, LONG TIME AGO

The third reason why Miller's argument misses the mark is actually quite understandable. It arises from the confusion of two separate ideas—the theory that life was intelligently designed and the theory that the earth is young. Because religious groups who strongly advocate both ideas have been in the headlines over the past several decades, much of the public thinks that the two ideas are necessarily linked. Implicit in Ken Miller's argument about pseudogenes, and absolutely required for his conclusions, is the idea that the designer had to have made life recently. *That is not a part of intelligent-design theory.* The conclusion that some features of life were designed can be made in the absence of knowledge about when the designing took place. A child who looks at the faces on Mt. Rushmore immediately knows that they were designed but might have no idea of their history; for all she knows, the faces might have been designed the day before she got there, or might have been there since the beginning of time. An art museum might display a statue of a bronze cat purportedly made in Egypt thousands of years ago—until the statue is examined by technologically advanced methods and shown to be a modern forgery. In either case, though, the bronze cat was certainly designed by an intelligent agent.

The irreducibly complex biochemical systems that I have discussed in this book did not have to be produced recently. It is entirely possible, based simply on an examination of the systems themselves, that they were designed billions of years ago and that they have been passed down to the present by the normal processes of cellular reproduction. Perhaps a speculative scenario will illustrate the point. Suppose that nearly four billion years ago the designer made the first cell, already containing all of the irreducibly complex biochemical systems discussed here and many others. (One can postulate that the designs for systems that were to be used later, such as blood clotting, were present but not "turned on." In present-day organisms plenty of genes are turned off for a while, sometimes for generations, to be turned on at a later time.) Additionally, suppose the designer placed into the cell some other systems for which we cannot adduce enough evidence to conclude design. The cell containing the designed systems then was left on autopilot to reproduce, mutate, eat and be eaten, bump against rocks, and suffer all the vagaries of life on earth. During this process, *pace* Ken Miller, pseudogenes might occasionally arise and a complex organ might become nonfunctional. These chance events do not mean that the initial biochemical systems were not designed. The cellular warts and wrinkles that Miller takes as evidence of evolution may simply be evidence of age.

Simple ideas can take a surprising amount of time to be properly developed. One way in which a simple idea can be sidetracked is through conflation with an extraneous idea. When it is considered by itself—away from logically unrelated ideas—the theory of intelligent design is seen to be quite robust, easily answering the argument from imperfection.

A COMPLICATED WORLD

The production of some biological improvements by mutation and natural selection—by evolution—is quite compatible with intelligent-design theory. Stephen Jay Gould of Harvard University has made much of the panda's "thumb." The giant panda lives on a diet of bamboo. To strip the leaves off bamboo shoots the panda grips them in its paw with a bony protuberance that emanates from its wrist; the normal five digits are also present. Gould argues that a designer would have given the panda a real opposable thumb, and so he concludes that the panda's thumb evolved. Gould's conclusion, though, suffers from the problems I have discussed earlier. He assumes the designer would act as he would, that pandas' thumbs "ought" to be arranged a different way. He then takes those assertions to be positive evidence for evolution. Gould has never done the science to support his idea: he has not shown or calculated what the minimum extension of the wristbone would have to be to help the panda; he has not justified the behavioral changes that would be necessary to take advantage of the change in bone structure; and he has not mentioned how pandas ate before acquiring the thumb. He has not done anything except to spin a tale.

But let's ignore those questions for now; let's assume that the story actually happened. Even then, why is Gould's panda scenario incompatible with intelligent-design theory? The panda's thumb is a black box. It is *entirely* possible that in the production of the panda's thumb, no new irreducibly complex systems were required in the cell. It is possible that the systems that were already present—the systems that make muscle proteins and nerve fibers, that lay down bone and matrix protein, that cause cells to divide for a while and then cease division—were enough. It is possible that these systems were quite sufficient to cause a bone protuberance when some chance event perturbed their normal pattern of operation, and it is possible that natural selection then favored this change. Design theory has nothing to say about a biochemical or biological system unless all the components of the system are known and it is demonstrated that the system is composed of several interacting parts. Intelligent-design theory can coexist quite peacefully with the panda's thumb.

We live in a complex world where lots of different things can happen. When deciding how various rocks came to be shaped the way they are, a geologist might consider a whole range of factors: rain, wind, the movement of glaciers, the activity of moss and lichens, volcanic action, nuclear explosions, asteroid impact, or the hand of a sculptor. The shape of one rock might have been determined primarily by one mechanism, the shape of another rock by another mechanism. The possibility of a meteor's impact does not mean that volcanos can be ignored; the existence of sculptors does not mean that many rocks are not shaped by weather. Similarly, evolutionary biologists have recognized that a number of factors might have affected the development of life: common descent, natural selection, migration, population size, founder effects (effects that may be due to the limited number of organisms that begin a new species), genetic drift (the spread of "neutral," nonselective mutations), gene flow (the incorporation of genes into a population from a separate population), linkage (occurrence of two genes on the same chromosome), meiotic drive (the preferential selection during sex cell production of one of the two copies of a gene inherited from an organism's parents), transposition (the transfer of a gene between widely separated species by nonsexual means), and much more. The fact that some biochemical systems may have been designed by an intelligent agent does not mean that any of the other factors are not operative, common, or important.

WHAT WILL SCIENCE DO?

The discovery of design expands the number of factors that must be considered by science when trying to explain life. What will be the effect of the awareness of intelligent design on different branches of science? Biologists who are working at the cellular level or above can continue their research without paying much attention to design, because above the cellular level organisms are black boxes, and design is difficult to prove. So those who labor in the fields of paleontology, comparative anatomy, population genetics, and biogeography should not invoke design until the molecular sciences show that design has an effect at those higher levels. Of course, the possibility of design should cause researchers in biology to hesitate before claiming that a particular biological feature has been produced substantially by another mechanism, such as natural selection or transposition. Instead, detailed models should be produced to justify the assertion that a given mechanism produced a given biological feature.

Unlike Darwinian evolution, the theory of intelligent design is new to modern science, so there are a host of questions that need to be answered and much work lies ahead. For those who work at the molecular level, the challenge will be to rigorously determine which systems were designed and which might have arisen by other mechanisms. To reach a conclusion of design will require the identification of the components of an interacting molecular system and the roles they play, as well as a determination that the system is not a composite of several separable systems. To reach a strong presumption of nondesign will require the demonstration that a system is not irreducibly complex or does not have much specificity between its components. To decide borderline cases of design will require the experimental or theoretical exploration of models whereby a system might have developed in a continuous manner, or a demonstration of points where the development of the system would necessarily be discontinuous.

Future research could take several directions. Work could be undertaken to determine whether information for designed systems could lie dormant for long periods of time, or whether the information would have to be added close to the time when the system became operational. Since the simplest possible design scenario posits a single cell—formed billions of years ago—that already contained all information to produce descendant organisms, other studies could test this scenario by attempting to calculate how much DNA would be required to code the information (keeping in mind that much of the information might be implicit). If DNA alone is insufficient, studies could be initiated to see if information could be stored in the cell in other ways—for example, as positional information. Other work could focus on whether larger, compound systems (containing two or more irreducibly complex systems) could have developed gradually or whether there are compounded irreducibilities

The preceding are just the obvious questions that flow from a theory of design. Undoubtedly, more and better-formed questions will be generated as more and more scientists grow curious about design. The theory of intelligent design promises to reinvigorate a field of science grown stale from a lack of viable solutions to dead-end problems. The intellectual competition created by the discovery of design will bring sharper analysis to the professional scientific literature and will require that assertions be backed by hard data. The theory will spark experimental approaches and new hypotheses that would otherwise be untried. A rigorous theory of intelligent design will be a useful tool for the advancement of science in an area that has been moribund for decades.

CHAPTER 11

SCIENCE, PHILOSOPHY, RELIGION

THE DILEMMA

Over the past four decades modern biochemistry has uncovered the secrets of the cell. The progress has been hard won. It has required tens of thousands of people to dedicate the better parts of their lives to the tedious work of the laboratory. Graduate students in untied tennis shoes scraping around the lab late on Saturday night; postdoctoral associates working fourteen hours a day seven days a week; professors ignoring their children in order to polish and repolish grant proposals, hoping to shake a little money loose from politicians with larger constituencies to feed—these are the people that make scientific research move forward. The knowledge we now have of life at the molecular level has been stitched together from innumerable experiments in which proteins were purified, genes cloned, electron micrographs taken, cells cultured, structures determined, sequences compared, parameters varied, and controls done. Papers were published, results checked, reviews written, blind alleys searched, and new leads fleshed out.

The result of these cumulative efforts to investigate the cell—to investigate life at the molecular level—is a loud, clear, piercing cry of *"design!"* The result is so unambiguous and so significant that it must be ranked as one of the greatest achievements in the history of science. The discovery rivals those of Newton and Einstein, Lavoisier and Schrödinger, Pasteur, and Darwin. The observation of the intelligent design of life is as momentous as the observation that the earth goes around the sun or that disease is caused by bacteria or that radiation is emitted in quanta. The magnitude of the victory, gained at such great cost through sustained effort over the course of decades, would be expected to send champagne corks flying in labs around the world. This triumph of science should evoke cries of "Eureka!" from ten thousand throats, should occasion much hand-slapping and high-fiving, and perhaps even be an excuse to take a day off.

But no bottles have been uncorked, no hands slapped. Instead, a curious, embarrassed silence surrounds the stark complexity of the cell. When the subject comes up in public, feet start to shuffle, and breathing gets a bit labored. In private people are a bit more relaxed; many explicitly admit the obvious but then stare at the ground, shake their heads, and let it go at that.

Why does the scientific community not greedily embrace its startling discovery? Why is the observation of design handled with intellectual gloves? The dilemma is that while one side of the elephant is labeled intelligent design, the other side might be labeled God.

A non-scientist might ask the obvious question: so what? The idea that a being such as God exists is not unpopular—far from it. Polls show that more than 90 percent of Americans believe in God, and that about half attend religious services regularly. Politicians invoke the name of God with great regularity (more often around election time). Many football coaches pray with their teams before games, musicians compose hymns, artists paint pictures of religious events, organizations of businessmen gather for prayers. Hospitals and airports have chapels; the army and Congress employ chaplains. As a country we honor people, such as Martin Luther King, whose actions were deeply rooted in a belief in God. With all of this public affirmation, why should science find it difficult to accept a theory that supports what most people believe anyway? There are several reasons. The first is a problem that many of us are prone to—simple chauvinism. The other reasons depend on historical and philosophical relationships that are peculiar to science. These various reasons all interact with one another in complex ways, but let's try to tease them apart.

ALLEGIANCE

People who dedicate their lives to a noble pursuit often become fiercely loyal to it. For example, a college president may devote all her efforts to strengthening her school, because educating people is an estimable service. A career army officer will work to improve his branch of the service, because defending one's country is a worthy purpose. Sometimes, however, loyalty to a particular institution causes a conflict of interest with the purpose the institution serves. The officer might rush his troops into battle so that the army will be credited with victory, even though it might be prudent to let the air force see the first action. The college president might persuade her state's congressmen to earmark federal money for a new building on her campus, even though the money might serve education better elsewhere.

Science is a noble pursuit that can engender fierce loyalty. The purpose of science is to explain the physical world—a very serious enterprise. However, other academic disciplines (principally philosophy and theology) also are in the business of explaining parts of the world. Although most of the time these disciplines stay out of each other's way, sometimes they conflict. When that happens some dedicated people put their discipline ahead of the goal it is supposed to serve.

A good example of disciplinary chauvinism can be seen in Robert Shapiro's fine book, *Origins: A Skeptic's Guide to the Creation of Life on Earth*. After presenting a very readable, very devastating critique of scientific studies on the origin of life, Shapiro proclaims his steadfast loyalty—not to the goal of "explaining the physical world," but to science:

> Some future day may yet arrive when all reasonable chemical experiments run to discover a probable origin for life have failed unequivocally. Further, new geological evidence may indicate a sudden appearance of life on the earth. Finally, we may have explored the universe and found no trace of life, or process leading to life, elsewhere. In such a case, some scientists might choose to turn to religion for an answer. Others, however, myself included, would attempt to sort out the surviving less probable scientific explanations in the hope of selecting one that was still more likely than the remainder.[1]

Shapiro goes on gamely to say that things don't look quite so bleak right now, pretty much contradicting everything he had written to that point. He can rest secure in the knowledge that there will never be a time when all experiments have "failed unequivocally," just as there will never be a time when the existence of the Loch Ness Monster has been absolutely ruled out. And the time when the universe will have been fully explored is comfortably far off.

Now, a nonpartisan might think that if none of the most likely scientific hypotheses panned out, then maybe a fundamentally different explanation is called for. After all, the origin of life was an historical event—not like, say, the search for a cancer cure, where science can keep trying till it succeeds. Maybe the origin of life just didn't happen by undirected chemical reactions, as Shapiro hopes. To an active participant in the search, however, a conclusion of design can be deeply unsatisfying. The thought that knowledge of the mechanisms used to produce life might be forever beyond their reach is admittedly frustrating to many scientists. Nonetheless, we must be careful not to allow distate for a theory to prejudice us against a fair reading of the data.

Loyalty to an institution is fine, but bare loyalty is not an argument. All in all, the effect of scientific chauvinism on theories of the development of life is an important sociological artifact to consider, but ultimately its intellectual importance for the topic of intelligent design is nil.

HISTORY LESSON

The second reason for the reluctance of science to deal with the elephant comes from history. From the time it was first proposed, some scientists have clashed with some theologians over Darwin's theory of evolution. Although many scientists and theologians thought that Darwinian evolution could be reconciled rather easily with the basic beliefs of most religions, publicity always focuses on conflict. The tone was probably set for good when Anglican bishop Samuel Wilberforce debated Thomas Henry Huxley, a scientist and strong advocate of evolution, about a year after Darwin's seminal book was published. It was reported that the bishop—a good theologian but poor biologist—ended his speech by asking, "I beg to know, is it through his grandfather or grandmother that Huxley claims his descent from a monkey?" Huxley muttered something like, "The Lord has delivered him into my hands," and proceeded to give the audience and the bishop an erudite biology lesson. At the end of his exposition Huxley declared that he didn't know whether it was through his grandmother or grandfather that he was related to an ape, but that he would rather be descended from simians than be a man possessed of the gift of reason and see it used as the bishop had used it that day. Ladies fainted, scientists cheered, and reporters ran to print the headline: "War Between Science and Theology."

The event in America that defined the public perception of the relationship of science to theology was the Scopes trial. In 1925 John Scopes, a high school biology teacher in the tiny town of Dayton, Tennessee, volunteered to be arrested for violating a previously unenforced state law forbidding the teaching of evolution. The involvement of high-profile lawyer Clarence Darrow for the defense and three-time losing presidential candidate William Jennings Bryan for the prosecution guaranteed the media circus that ensued. Although Scopes's team lost at trial, his conviction was overturned on a technicality. More importantly, the publicity set a tone of antagonism between religion and science.

The Scopes trial and the Huxley-Wilberforce debate happened long ago, but more recent events have kept the conflict simmering. Over the past several decades groups that, for religious reasons, believe that the earth is relatively young (on the order of ten thousand years) have tried to have their viewpoint taught to their children in public schools. The sociological and political factors involved in the situation are quite complex—a powerful mix of such potentially divisive topics as religious freedom, parental rights, government control of education, and state versus federal rights—and are made all the more emotional because the fight is over children.

Because the age of the earth can be inferred from physical measurements, many scientists quite naturally felt that the religious groups had entered their area of expertise and called them to account. When the groups offered physical evidence that they said supported a young earth, scientists hooted it down as incompetent and biased. Tempers flared on both sides, and much ill will was built up. Some of the ill will has been institutionalized; for example, an organization called the National Center for Science Education was set up a dozen years ago—when several states were passing laws congenial to creationism—to battle creationists whenever they try to influence public school policy.

These conflicts reverberate into the present. In 1990 *Scientific American* asked a science writer named Forrest Mims to write several columns for the "Amateur Scientist" feature of their magazine. "Amateur Scientist" treats topics such as measuring the length of lightning bolts, building portable solar observatories, and making a home seismometer to record earth movements—fun projects for those whose hobby is science. The understanding was that if the editors and readers liked the columns, Mims would be hired as a permanent writer. The trial columns all went very well, but when Mims came to New York for a final interview he was asked if he believed in evolution. Mims replied, well, no, he believed in the biblical account of creation.

The magazine refused to hire him. *Scientific American* was afraid that merely having a creationist on the staff would hurt its reputation among scientists, even though Mims was well qualified and had no plans to write about evolution. Undoubtedly scenes from *Inherit the Wind* (the movie based loosely on the Scopes trial) and news clips of battles between creationists and their political foes flickered through the minds of the magazine's editors. Such widely reported mini-conflicts as the Mims affair—even though they have nothing directly to do with the real intellectual issues about how life on earth came to be—fuel the historical flames of conflict between science and religion, and persuade many people that you must belong to one camp or to the other.

The historical events in which scientists have clashed with religious groups are real and cause real emotional reactions. They make some well-meaning people think that a demilitarized zone should be maintained between the two, with no fraternization allowed. Like scientific chauvinism, however, the importance of the historical clashes for actual scientific understanding of the development of life is essentially zero. I am not naively hoping that biochemistry's discoveries can be evaluated free from the shadows of history, but to the greatest extent possible, they should be.

Unlike chauvinist and historical arguments, philosophical arguments that seek to head off a conclusion of intelligent design are substantive; they affect the issues on an intellectual level, not just an emotional one. There are several different philosophical issues. Let's examine them.

THE RULE

Richard Dickerson is a prominent biochemist, an elected member of the elite National Academy of Sciences who specializes in X-ray crystallographic studies of proteins and DNA. He and the workers in his laboratory have made notable contributions to our understanding of the structure of the molecules of life. He is not the most prominent scientist in the United States, and his contributions have not been the flashiest, but Dickerson is in many ways the paradigm of a dedicated scientist. He is the sort of person, and his the sort of professional situation, that thousands of graduate students have in mind as they labor day and night in the laboratory, dreaming of the day when they too will be respected members of the scientific community.

Dickerson's published opinions nicely capture the way many scientists view the world of religion. A few years ago Dickerson wrote a short essay summarizing his views on science vis-à-vis religion and had the essay published in both the *Journal of Molecular Evolution* (a secular scientific journal) and *Perspectives on Science and Christian Faith* (a journal published by the American Scientific Affiliation, which is an organization of scientists who are also evangelical Christians). So it is safe to conclude that Dickerson was not just directing his remarks to people who already shared his ideas—he was making an honest effort to present what he thought were reasonable and convincing views to persons with diverging opinions. Because of its consonance with most scientists' view of science, Dickerson's essay makes a useful springboard for considering how the theory of intelligent design fits into science:

> Science, fundamentally, is a game. It is a game with one overriding and defining rule:
>
> *Rule No. 1:* Let us see how far and to what extent we can explain the behavior of the physical and material universe in terms of purely physical and material causes, without invoking the supernatural.
>
> Operational science takes no position about the existence or nonexistence of the supernatural; it only requires that this factor is not to be invoked in scientific explanations. Calling down special-purpose miracles as explanations constitutes a form of intellectual "cheating." A chess player is perfectly capable of removing his opponent's King physically from the board and smashing it in the midst of a tournament. But this would not

make him a chess champion, because the rules had not been followed. A runner may be tempted to take a short-cut across the infield of an oval track in order to cross the finish line ahead of his faster colleague. But he refrains from doing so, as this would not constitute "winning" under the rules of the sport.[2]

Let's rephrase Dickerson's rule to the following: Science must invoke only natural causes, and explain by reference only to natural law.[3] The reformulation makes explicit what is strongly implied by the phrase "let us see how far."

In his essay, then, Dickerson does not say scientific evidence has shown that the supernatural has never affected nature (for those concerned about the definition of *supernatural*, substitute "higher intelligence"). Rather, he argues that in principle, science should not invoke it. The clear implication is that it should not be invoked *whether it is true or not*. It is relevant to our evaluation of his argument that Dickerson is a member of the American Scientific Affiliation, so he believes in God. He has no a priori reason to think that nothing beyond nature exists, but he thinks it is not good science to offer the supernatural as an explanation for a natural event.

(Incidentally, scientists who believe in God or a reality beyond nature are much more common than popular media stories lead one to believe—there is no reason to think that the figure of 90 percent of the general population that believes in God is much different for scientists. Ken Miller, whose argument from imperfection I analyzed in the last chapter, is like myself a Roman Catholic, and he makes the point in public talks that belief in evolution is quite compatible with his religious views. I agree with him that they are compatible.[4] The compatibility or lack of compatibility, however, is irrelevant to the scientific question of whether Darwinian evolution of biochemical systems is true.)

It is important to note that Dickerson's argument is not itself a scientific one—it was not discovered by an experiment in a laboratory; it is not the result of mixing chemicals in a test tube; it is not a testable hypothesis. Rather, the argument is philosophy. It may be good philosophy, or it may not. Let's examine it more closely.

Most people would be surprised to learn that "science, fundamentally, is a game." Certainly the taxpayers who fund science to the tune of several tens of billions of dollars a year would be surprised. They probably think they're spending their money to find cures and treatments for cancer, AIDS, and heart disease. Citizens concerned about diseases they have or may acquire in old age want science to be able to cure the disease, not to play a game that has no bearing on reality. I doubt that Darwin or Newton or Einstein thought of science this way. The giants of science were motivated by a thirst to know the real world, and some (such as Galileo) paid a price for their knowledge. For students, science textbooks do not present science as a game but as a noble search for truth. Most people, from ordinary taxpayers to prominent scientists, would more likely view science not as a game but as *a vigorous attempt to make true statements about the physical world*.

The assertion that science is a game does not stand up to even a cursory examination. No one would seriously maintain that position very long if questioned about it. Richard Dickerson himself would probably quickly retract his statement if he had to defend it in front of a skeptical audience. Clearly Dickerson has something else in mind. Perhaps he means that science, like games, is a rule-bound activity. Other serious activities, like criminal trials and political campaigns, are rule-bound activities. Is science also? If so, what are the rules?

Let's focus on the second question. Dickerson mentions just one rule, the one disbarring the supernatural. Where did he get it? Is it written in a textbook? Is it found in the by-laws of scientific societies? No, of course not. You can scan all the textbooks that are used for science instruction in all of the major universities of this country, and you will not see the "one overriding and defining rule." Nor will you see any other general rules proscribing how science is to be conducted (other than safety rules, exhortations to honesty, and the like).

Nonetheless, let us ask, how does Dickerson's rule help? Does the rule tell us what subjects are beyond science's competence? Does it give us guidelines for discriminating science from pseudo science? Does it even give a definition of what science is? The answer to all of these questions is no. A few years ago an article by a Nobel laureate was published in a prestigious scientific journal; the article analyzed the rationality of people who forgo having children in order to help others (such as, say, Mother Teresa) in terms of evolutionary reproductive strategies.[5] Such "science" does not violate Dickerson's rule. Dickerson's "one overriding and defining rule" would happily tolerate the discredited nineteenth-century science of phrenology (the attempt to discern the intelligence and character of people from the shape of their skulls). His rule gives us no guidance about the legitimacy of Marxism and Freudianism, the "sciences" of history and the mind, respectively. The rule does not help us decide in advance if putting leeches on sick people or bleeding them to reduce their fever will work. So it seems that many things could claim the title of "science" under Dickerson's rule, as long as they invoke only material forces, however vague and elusive.

In fact, Dickerson's rule is more like a professional aphorism—like "the customer is always right" or "location, location, location." It is what the old professionals have lived by, what they think works, and it encapsulates some of the wisdom that they wish to pass on to the younger professional generation. Behind Dickerson's rule are vague images of Vikings attributing thunder and lightning to the work of the gods, and of witch doctors trying to drive out evil spirits from sick people. Closer to modern science are memories of Isaac Newton himself proposing that God occasionally intervened to stabilize the solar system. The anxiety is that if the supernatural were allowed as an explanation, then there would be no stopping it—it would be invoked frequently to explain many things that in reality have natural explanations. Is this a reasonable fear?

No one can predict the behavior of human beings, but it seems to me that the fear of the supernatural popping up everywhere in science is vastly overblown. If my graduate student came into my office and said that the angel of death killed her bacterial culture, I would be disinclined to believe her. The *Journal of Biological Chemistry* is unlikely to start a new section on the spiritual regulation of enzyme activity. Science has learned over the past half millennium that the universe operates with great regularity the great majority of the time, and that simple laws and predictable behavior explain most physical phenomena. Historians of science have emphasized that science was born from a religious culture—Europe in the Middle Ages—whose religious traditions included a rational God who made a rational, understandable, law-bound universe.[6] Both science *and* religion expect that the world will almost always spin according to the fixed law of gravity.

There are, of course, exceptions. Sometimes unique historical events must be invoked to explain an effect. The fossil record shows that about 60 million years ago, the dinosaurs all died out within a geologically brief time period. One theory offered to explain this is that a large meteor crashed into the earth, sending clouds of dust high into the atmosphere and perhaps causing many plants to die, disrupting the food chain. Some indirect evidence supports the hypothesis—levels of the element iridium, rarely found on earth but more frequent in meteors, are elevated in rocks from that time period. The hypothesis has been accepted by many scientists. Nonetheless, there has not been a rush to postulate meteors as the cause of all sorts of things. No one has said that meteors caused the Grand Canyon, or the extinction of horses in North America. No one has said that the dust of tiny, invisible meteorites causes asthma, or that meteorites initiate tornados. The hypothesis of the involvement of a meteor in the extinction of the dinosaurs was evaluated on the basis of the physical evidence for the particular historical event. There is every reason to expect that evidence will be evaluated on a case-by-case basis if meteors are invoked to explain other historical events.

Similarly, hypotheses for the involvement of an intelligent agent in the development of life or other historical events have to be evaluated on a case-by-case basis. As noted in Chapter 9, the evidence is overwhelming for some biochemical systems, undetectable for others. If a scientist postulates the involvement of intelligence in some other event, then the onus will be on him or her to support that assertion with observable evidence. The scientific community is not so frail that its healthy skepticism will turn into gullibility.

Another concern that might lie behind Dickerson's essay is for the "scientific method." Hypothesis, careful testing, replicability—all these have served science well. But how can an intelligent designer be tested? Can a designer be put in a test tube? No, of course not. But neither can extinct common ancestors be put in test tubes. The problem is that whenever science tries to explain a unique historical event, careful testing and replicability are by definition impossible. Science may be able to study the motion of modern comets, and test Newton's laws of motion that describe how the comets move. But science will never be able to study the comet that putatively struck the earth many millions of years ago. Science can, however, observe the comet's lingering *effects* on the modern earth. Similarly, science can see the effects that a designer has had on life.

The final point I wish to make about Richard Dickerson's argument is that although he certainly didn't intend it, it is a prescription for timidity. It tries to restrict science to more of the same, disallowing a fundamentally different explanation. It tries to place reality in a tidy box, but the universe will not be placed in a box. The origin of the universe and the development of life are the physical underpinnings that resulted in a worldful of conscious agents. There is no a priori reason to think that those bedrock events are to be explained in the same way as other physical events. Science is not a game, and scientists should follow the physical evidence wherever it leads, with no artificial restrictions.

GHOSTBUSTERS

The fourth and most powerful reason for science's reluctance to embrace a theory of intelligent design is also based on philosophical considerations. Many people, including many important and well-respected scientists, just don't *want* there to be anything beyond nature. They don't want a supernatural being to affect nature, no matter how brief or constructive the interaction may have been. In other words, like young-earth creationists, they bring an a priori philosophical commitment to their science that restricts what kinds of explanations they will accept about the physical world. Sometimes this leads to rather odd behavior.

It was only about seventy years ago that most scientists thought the universe was infinite in age and size. That view had been held by some Greek philosophers in antiquity, as well as by diverse religious groups, and by those who thought there was nothing beyond nature. In contrast Judaism and then Christianity thought that the universe was created in time and was not eternal. Having few scientists among them, the early Jews did not try to adduce evidence for the finiteness of the universe, and in the Middle Ages Thomas Aquinas, the eminent theologian, said that it could be known only through faith that the universe had a beginning. But time marches on. Earlier this century Einstein discovered that his general theory of relativity predicted an unstable universe—one that would either expand or contract, but would not remain stationary. Einstein was repulsed by such a universe and, in what he later admitted was the greatest mistake of his career, inserted a "correction factor" into his equations solely to make them predict a stationary, eternal universe.

As parents and teachers always say, cheaters never prosper. A short time later the astronomer Edwin Hubble observed that wherever in the sky he pointed his telescope, the stars appeared to be moving away from the earth. (He couldn't actually see the stars moving. Rather, he inferred their motion from a phenomenon called a "Doppler shift," in which stars that move away from an observer emit light of a slightly longer wavelength—the faster they move, the greater the change in the wavelength.) Furthermore, the speed with which the stars were receding was proportional to their distance from the earth. This was the first observational evidence that Einstein's unfudged equations were correct in their prediction concerning the expansion of the universe. And it did not take a rocket scientist (although plenty were around) to mentally reverse the expanding universe and conclude that at some time in the past, all of the matter in the universe was concentrated into a very small space. This was the beginning of the Big Bang hypothesis.

To many the notion of the Big Bang was loaded with overtones of a supernatural event—the creation, the beginning of the universe. The prominent physicist A.S. Eddington probably spoke for many in voicing his utter disgust with such an idea:

> Philosophically, the notion of an abrupt beginning to the present order of Nature is repugnant to me, as I think it must be to most; and even those who would welcome a proof of the intervention of a Creator will probably consider that a single winding-up at some remote epoch is not really the kind of relation between God and his world that brings satisfaction to the mind.[7]

Nonetheless, despite its religious implications, the Big Bang was a scientific theory that flowed naturally from observational data, not from holy writings or transcendental visions. Most physicists adopted the Big Bang theory and set their research programs accordingly. A few, like Einstein before them, didn't like the extra-scientific implications of the theory and labored to develop alternatives.

In the middle part of the century the astronomer Fred Hoyle championed another theory of the universe, called the steady-state theory. Hoyle proposed that the universe was infinite and eternal, but he also admitted that the universe was expanding. Since a universe that has been expanding for an infinite period of time would become infinitely thinned out, even if it started with an infinite amount of matter, Hoyle had to explain why our present universe is relatively dense. The eminent scientist proposed that matter was continually coming into existence in outer space at the rate of about one hydrogen atom per cubic mile of space per year. Now, it must be emphasized that Hoyle was proposing creation of hydrogen *from nothing and with no cause*. The matter simply popped into existence at the required rate. Since he had no observational evidence to support this notion, why did Hoyle propose it? It turns out that Hoyle, like Eddington, thought that the Big Bang strongly implied the supernatural and found the prospect extremely distasteful.

Hoyle's steady-state theory always had a difficult time explaining much of the observational evidence from astronomy. In the 1960s the astronomers Penzias and Wilson finally put the theory out of its misery with their observation of background radiation. They saw that microwaves are bombarding the earth from every direction with an astonishing uniformity of intensity. Such background radiation was predicted to be an indirect result of the Big Bang. The observation of the background radiation was, and still is, taken to be the crowning glory of the Big Bang theory.

It is impossible to deny that the Big Bang has been an enormously fruitful physical model of the universe and, even though large questions remain (as they inevitably do in basic science), that the model was justified by the observational data. Scientists such as Einstein, Eddington, and Hoyle fudged and twisted in their efforts to resist a scientific theory that flowed naturally from the data because they thought they would be forced to accept unpleasant philosophical or theological conclusions. They weren't; they had other options.

DON'T FENCE ME IN

The success of the Big Bang model had nothing to do with its religious implications. It seemed to agree with the Judaeo-Christian dogma of a beginning to the universe; it seemed to disagree with other religions that believed the universe to be eternal. But the theory justified itself by reference to observational data—the expansion of the universe—and not by invoking sacred texts or the mystical experiences of holy men. The model came straight from the observational evidence; it was not fit to a Procrustean bed of religious dogma.

But it should also be noticed that the Big Bang, although friendly to a religious point of view, does not forcibly compel that belief. No person is required by dint of logic to reach any particular supernatural conclusion solely on the basis of scientific observations and theories. This is seen initially in Einstein's and Hoyle's attempt to come up with alternative models that would fit the observational data and avoid the unpleasant thought of a start to the universe. When the steady-state theory was finally discredited, other theories sprang up that would obviate the philosophical bind of an absolute beginning. The most popular option was the idea of a cycling universe, in which the expansion that started with the Big Bang would eventually slow down and, under the force of gravity, all matter would collapse again in a "Big Crunch." From there, the story goes, perhaps another Big Bang would occur, and endless repetitions of this cycle would recapture a nature that was infinite in time. It is interesting (though scientifically irrelevant) that the notion of a cycling universe would be compatible with many religions, including those of the ancient Egyptians, Aztecs, and Indians.[8]

The idea of a cycling universe seems to be out of favor in physics these days. Insufficient matter to cause a future gravitational collapse has been observed—and even if such matter existed, calculations show that successive cycles would become longer and longer, eventually ending with a non-contracting universe. But even if this option is discredited, other ideas are available to take the sting out of the Big Bang. A more recent proposal would have it that the actual universe is enormously larger than what we observe, and that the portion of the universe that we see is merely a "bubble" in an infinite universe. And physicist Stephen Hawking has proposed that although the universe is finite, it would not have a beginning if something in his mathematical equations that he calls "imaginary time" actually existed. Another idea is that infinitely many universes exist, and that the universe in which we find ourselves just happens to have the narrow conditions required for life. This idea was popularized under the name of the "anthropic principle." In essence the anthropic principle states that very many (or infinitely many) universes exist with varying physical laws, and that only the ones with conditions suitable for life will in fact produce life, perhaps including conscious observers. So perhaps a zillion barren universes exist somewhere; we live in the zillion and first universe because it has the physical properties that are compatible with life.

The anthropic principle strikes most people as plain silly, probably because they aren't quite sure where we would put all those other universes. Other ideas, however, are available for the person who still does not want to invoke the supernatural. In quantum physics it is believed that microscopic entities called "virtual particles" can pop into existence by borrowing energy from the surroundings (confusingly called the "vacuum," even though the word is not used by physicists to mean "nothing"). Some physicists have taken this idea just a bit further and proposed that the entire universe simply popped into existence, not from any surroundings, but from absolutely nothing—"a quantum fluctuation from nonbeing to being"—and without a cause. This shows how some scientists have learned to think big compared to the days when Fred Hoyle was modestly proposing the occasional uncaused creation of hydrogen atoms.

No experiment has been done to support the notion of bubble universes, imaginary time, or the zillion anthropic universes. Indeed, it seems that no experiment could detect them in principle. Since they or their effects cannot be observed, then they are metaphysical postulates, no more accessible to experimental investigation than an admittedly supernatural being. They do science no good. Their only use is as an escape hatch from the supernatural.

The point of the above discussion is that even though the Big Bang hypothesis may appear at first blush to support a particular religious idea, no scientific theory can compel belief in a positive religious tenet by sheer force of logic. Thus, to explain the universe a person can postulate unobservables, like the theory that there are infinitely many universes and the theory that ours is just a bubble in a larger universe. Or one can hold out the hope that theories that look implausible today, such as the steady-state theory or the theory of the oscillating universe, might look more plausible tomorrow when calculations are redone or new measurements are taken. Or one can simply abandon the principle of causation, as seen in theories that propose that the universe came into being uncaused. Most other people may regard the ideas as pretty giddy; nonetheless, they don't violate the observational evidence.

ALIENS AND TIME TRAVELERS

Saying that the universe began in a Big Bang is one thing, but saying life was designed by an intelligence is another. The phrase *Big Bang* itself evokes only images of an explosion, not necessarily a person. The phrase *intelligent design* seems more urgent and quickly invites questions about who the designer might have been. Will persons with philosophical commitments against the supernatural be painted into a corner by the theory? No. The human imagination is too powerful.

By any measure, Sir Francis H. C. Crick is a smart man. Over forty years ago, while still a graduate student at Cambridge University, Crick and James Watson used X-ray crystallographic data to deduce the double helical structure of DNA, an accomplishment for which they later received the Nobel prize. Crick went on to contribute to the elucidation of the genetic code and to pose provocative conceptual questions on the function of the brain. Well into his seventies, he is pushing science further and faster than most of us will at the pinnacle of our powers.

Francis Crick also thinks that life on earth may have begun when aliens from another planet sent a rocket ship containing spores to seed the earth. This is no idle thought; Crick first proposed it with chemist Leslie Orgel in 1973 in an article entitled "Directed Panspermia" in a professional science journal called *Icarus*. A decade later Crick wrote a book, *Life Itself,* reiterating the theory; in a 1992 interview in *Scientific American* on the eve of the publication of his latest book, Crick reaffirmed that he thinks the theory is reasonable.

The primary reason Crick subscribes to this unorthodox view is that he judges the undirected origin of life to be a virtually insurmountable obstacle, but he wants a naturalistic explanation. For our present purposes, the interesting part of Crick's idea is the role of the aliens, whom he has speculated sent space bacteria to earth. But he could with as much evidence say that the aliens actually designed the irreducibly complex biochemical systems of the life they sent here, and also designed the irreducibly complex systems that developed later. The only difference is a switch to the postulate that aliens constructed life, whereas Crick originally speculated that they just sent it here. It is not a very big leap, though, to say that a civilization capable of sending rocket ships to other planets is also likely to be capable of designing life—especially if the civilization has never been observed. Designing life, it could be pointed out, does not necessarily require supernatural abilities; rather, it requires a lot of intelligence. If a graduate student in an earthbound lab today can plan and make an artificial protein that can bind oxygen, then there is no logical barrier to thinking that an advanced civilization on another world might design an artificial cell from scratch.

This scenario still leaves open the question of who designed the designer—how did life originally originate? Is a philosophical naturalist now trapped? Again, no. The question of the design of the designer can be put off in several ways. It could be deflected by invoking unobserved entities: perhaps the original life is totally unlike ourselves, consisting of fluctuating electrical fields or gases; perhaps it does not require irreducibly complex structures to sustain it. Another possibility is time travel, which has been seriously proposed by professional physicists in recent years. *Scientific American* informed the readers of its March 1994 issue that

> far from being a logical absurdity . . . the theoretical possibility of taking such an excursion into one's earlier life is an inescapable consequence of fundamental physical principles.

Perhaps, then, biochemists in the future will send back cells to the early earth that contain the information for the irreducibly complex structures we observe today. In this scenario humans can be their own aliens, their own advanced civilization. Of course, time travel leads to apparent paradoxes (things like grandsons shooting grandfathers before their offspring are born), but at least some physicists are ready to accept them. Most people, like me, will find these scenarios entirely unsatisfactory, but they are available for those who wish to avoid unpleasant theological implications.

In *The Blind Watchmaker* Richard Dawkins tells his readers that even if a statue of the Virgin Mary waved to them, they should not conclude they had witnessed a miracle.[9] Perhaps all the atoms of the statue's arm just happened to move in the same direction at once—a low-probability event to be sure, but possible. Most people who saw a statue come to life would tell Dawkins that there are more things in heaven and earth than are dreamt of in his philosophy, but they couldn't make him join the Church of England.

LIVE AND LET LIVE

Nor should they try. In a very real sense, the separateness of the spheres of science versus philosophy and religion is as it should be. Every person has available the data of his or her senses and, for the most part, can agree with other people on what that data is. To a large extent people of different philosophical and theological bents can also agree on scientific theories, such as gravitation or plate tectonics or evolution, to organize the data (even if the theories are ultimately incorrect). But the fundamental philosophical principles that underlie reality and the theological principles, or lack of principles, that can be garnered from philosophy and historical experience are at root chosen by the individual. A man or woman must be free to search for the good, the true, and the beautiful.

Refusal to give others broad latitude for their defining beliefs has led time and again to disaster. Intolerance does not arise when I think that I have found the truth. Rather it comes about only when I think that, because I have found it, everyone else should agree with me. Richard Dawkins has written that anyone who denies evolution is either "ignorant, stupid or insane (or wicked—but I'd rather not consider that.)"[10] It isn't a big step from calling someone wicked to taking forceful measures to put an end to their wickedness. John Maddox, the editor of *Nature*, has written in his journal that "it may not be long before the practice of religion must be regarded as anti-science."[11] In his recent book *Darwin's Dangerous Idea*, philosopher Daniel Dennett compares religious believers—90 percent of the population—to wild animals who may have to be caged, and he says that parents should be prevented (presumably by coercion) from misinforming their children about the truth of evolution, which is so evident to him.[12] This is *not* a recipe for domestic tranquility. It is one thing to try to persuade someone by polemics; it is entirely different to propose to coerce those who disagree with you. As the weight of scientific evidence shifts dramatically, this point should be kept prominently in mind. Richard Dawkins has said that Darwin made it possible to be an "intellectually fulfilled atheist."[13] The failure of Darwin's theory on the molecular scale may cause him to feel less fulfilled, but no one should try to stop him from continuing his search.

The scientific community contains many excellent scientists who think that there is something beyond nature, and many excellent scientists who do not. How then will science "officially" treat the question of the identity of the designer? Will biochemistry textbooks have to be written with explicit statements that "God did it"? No. The question of the identity of the designer will simply be ignored by science. The history of science is replete with examples of basic-but-difficult questions being put on the back burner. For example, Newton declined to explain what caused gravity, Darwin offered no explanation for the origin of vision or life, Maxwell refused to specify a medium for light waves once the ether was debunked, and cosmologists in general have ignored the question of what caused the Big Bang. Although the fact of design is easily seen in the biochemistry of the cell, identifying the designer by scientific methods might be extremely difficult. In the same way, Newton could easily observe gravity, but specifying its cause lay centuries in the future. When a question is too difficult for science to deal with immediately, it is happily forgotten while other, more accessible questions are investigated. If philosophy and theology want to take a crack at the question in the meantime, we scientists should wish them well, but reserve the right to jump back into the conversation when science has something more to add.

CURIOUSER AND CURIOUSER

The reluctance of science to embrace the conclusion of intelligent design that its long, hard labors have made manifest has no justifiable foundation. Scientific chauvinism is an understandable emotion, but it should not be allowed to affect serious intellectual issues. The history of skirmishes between religion and science is regrettable and has caused bad feelings all around. Inherited anger, however, is no basis for making scientific judgments. The philosophical argument (made by some theists) that science should avoid theories which smack of the supernatural is an artificial restriction on science. Their fear that supernatural explanations would overwhelm science is unfounded. Further, the example of the Big Bang theory shows that scientific theories with supernatural ramifications can be quite fruitful. The philo-sophical commitment of some people to the principle that nothing be-yond nature exists should not be allowed to interfere with a theory that flows naturally from observable scientific data. The rights of those people to avoid a supernatural conclusion should be scrupulously re-spected, but their aversion should not be determinative.

As we reach the end of this book, we are left with no substantive de-fense against what feels to be a strange conclusion: that life was de-signed by an intelligent agent. In a way, though, all of the progress of science over the last several hundred years has been a steady march to-ward the strange. People up until the Middle Ages lived in a natural world. The stable earth was at the center of things; the sun, moon, and stars circled endlessly to give light by day and night; the same plants and animals had been known since antiquity; kings ruled by divine right. Surprises were few.

Then it was proposed, absurdly, that the earth itself moved, spin-ning while it circled the sun. No one could feel the earth spinning; no one could see it. But spin it did. From our modern vantage, it's hard to realize what an assault on the senses was perpetrated by Copernicus and Galileo; they said in effect that people could no longer rely on even the evidence of their eyes.

Things got steadily worse over the years. With the discovery of fos-sils it became apparent that the familiar animals of field and forest had not always been on earth; the world had once been inhabited by huge, alien creatures who were now gone. Sometime later Darwin shook the world by arguing that the familiar biota was *derived* from the bizarre, vanished life over lengths of time incomprehensible to human minds. Einstein told us that space is curved and time is relative. Modern physics says that solid objects are mostly space, that subatomic parti-cles have no definite position, that the universe had a beginning.

Now it's the turn of the fundamental science of life, modern bio-chemistry, to disturb. The simplicity that was once expected to be the foundation of life has proven to be a phantom; instead, systems of horrendous, irreducible complexity inhabit the cell. The resulting real-ization that life was designed by an intelligence is a shock to us in the twentieth century who have gotten used to thinking of life as the result of simple natural laws. But other centuries have had their shocks, and there is no reason to suppose that we should escape them. Humanity has endured as the center of the heavens moved from the earth to beyond the sun, as the history of life expanded to encompass long-dead reptiles, as the eternal universe proved mortal. We will endure the opening of Darwin's black box.

THE CHEMISTRY OF LIFE

This appendix will give the interested reader an overview of biochemical principles that undergird life. It is not necessary to read the appendix to follow the arguments in the book, but it will set those arguments within a larger framework. Here I will discuss cells and the structures of several major classes of biomolecules—proteins and nucleic acids and, briefly, lipids and carbohydrates. I will then focus on the question of how genetic information is expressed and propagated. Of course, in such a short space the description must be sketchy, so I urge those who become intrigued by the mechanisms of life to borrow an introductory biochemistry text from the library. A fascinating Lilliputian world awaits.

CELLS AND MEMBRANES

The human body is composed of hundreds of trillions of cells. Other large animals and plants also are conglomerations of enormous numbers of cells. As the size of an organism decreases, however, the number of cells decreases also; for example, the small worm *C. elegans* contains only about a thousand cells. As we travel down the size scale we ultimately reach the unicellular phyla, such as yeast and bacteria. No independent life occurs below this level.

Examination of its structure shows why the cell is the fundamental unit of life. The defining feature of a cell is a membrane—a chemical structure that divides the outside world from the interior of the cell. With the protection of a membrane, a cell can maintain different conditions inside than prevails outside. For example, cells can concentrate nutrients in their interior so that they are available for energy production, and can prevent newly made structural materials from being washed away. In the absence of a membrane, the large array of metabolic reactions necessary to sustain life would quickly dissipate.

Cell membranes are made from amphiphilic molecules that are similar in ways to the soaps and detergents used in household cleaning. The word *amphiphilic* is from the Greek meaning "loves both"; an amphiphilic molecule "loves" two different environments: oil and water. The shape of the molecules is roughly similar to a lollipop with two sticks coming out the same side of the candy ball. The sticks usually consist of hydrocarbons (made from atoms of carbon and hydrogen) and, like other hydrocarbons such as gasoline, do not mix well with water. This is the oil-loving part of the molecule. Such regions of molecules are called *hydrophobic,* from the Greek for "water-fearing." The ball of the lollipop molecule, in contrast, generally has a chemical group that, like table salt or sugar, positively enjoys being in water. Such regions are called *hydrophilic* ("water-loving"). The two opposite parts of membrane molecules are chemically tied together and, like Siamese twins, must travel together despite dissimilar properties. But if one part of the molecule wants to be in water and the other part wants to be out of water, where does the molecule settle down?

Amphiphilic molecules solve their dilemma by associating with other amphiphilic molecules. When a large number of amphiphiles associate, the hydrophobic tails all huddle together to exclude water while the hydrophilic heads touch the water. An efficient way for the tails to be shielded from water while still allowing the water-loving groups access to water is to form two sheets (Figure A–1), called a *lipid bilayer.* If the two sheets remained flat, however, the hydrocarbons at the edges of the sheets would remain exposed to water. So the sheets close up, like a soap bubble.

Since the middle of the membrane bilayer is oily, many molecules

FIGURE A–1

A SEGMENT OF A LIPID BILAYER.

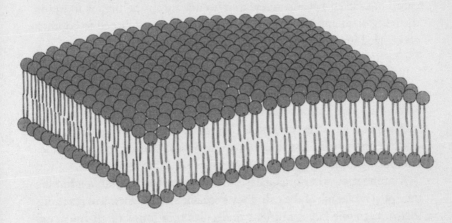

that strongly prefer a watery environment (such as salts and sugars) cannot cross the membrane. Thus we have a structure with an enclosed interior that can be different from the outside environment— the first step in making a cell.

The living world contains two fundamentally different type of cells: the *eukaryotes,* in which a second membrane, different from the cell membrane, encloses the nucleus of the cell; and the *prokaryotes,* which do not have this feature.[1] Prokaryotic organisms are invariably unicellular and are, in many ways, much simpler than eukaryotes.

Besides the cell membrane only a few features stand out in photographs of prokaryotes.[2] One is the *nucleoid,* the mass of cellular DNA (deoxyribonucleic acid) resting comfortably in the middle of the *cytoplasm* (the soluble cell contents). In addition to a membrane, prokaryotes have a second structure surrounding the cell, called the *cell wall.* Unlike the membrane, the cell wall is made of polysaccharide that is rigid and freely permeable to nutrients and small molecules. It confers mechanical strength, preventing the cell from rupturing under stress. Several structures stick out from the membrane of many prokaryotic cells. The function of the hairlike *pili* is largely unknown. The bacterial *flagellum* is used for locomotion; flagella rotate rapidly like a propeller to move the prokaryote along.

The second category of cells is the eukaryotes, which compose all multicellular organisms, as well as some single-celled organisms like yeast. Eukaryotic cells contain a number of subcellular spaces that are separated from the cytoplasm by their own membranes; these are called *organelles,* because they are reminiscent of the organs found in the body of an animal. Organelles allow the eukaryotic cell to conduct specialized functions in specialized compartments.

The first specialized organelle is the *nucleus,* which contains the cell's DNA. The membrane surrounding the nucleus is a highly specialized structure, perforated by large, eight-sided holes called *nuclear pores.* The pores are not passive punctures, however; they are active gatekeepers. No large molecule (like proteins or RNA) gets past the nuclear pores without the correct "password." This keeps molecules that belong in the cytoplasm out of the nucleus, and vice versa.

A number of other organelles stud the cytoplasm. *Mitochondria* are the "power plants" of the cell; they specialize in the chemical reactions that turn calorie-laden nutrient molecules into forms of chemical energy that the cell can use directly. Mitochondria have two membranes. The controlled "burning" of nutrient molecules generates a difference between the acidity of the space enclosed by the inner membrane and that enclosed between the inner and outer membranes. The controlled flow of acid between the two compartments generates energy, like the flow of water over a dam generates electrical power.

Lysosomes are small organelles bounded by a single membrane; essentially, they are bags of enzymes which degrade molecules that have outlived their usefulness. Molecules destined to be degraded in the lysosomes are transported there in small, coated vesicles (see Chapter 5). The acidity in the lysosome is one hundred to one thousand times greater than that in the cytoplasm. The increased acidity makes tightly folded proteins open up, and the open structures are then easily attacked by degradative enzymes.

The *endoplasmic reticulum* (ER) is an extensive, flattened, convoluted membrane system that is divided into two different components: the rough ER and the smooth ER. The rough ER gets its craggy appearance from numerous *ribosomes* attached to it; ribosomes are the cellular machinery that synthesize proteins. The smooth ER synthesizes lipids—fatty molecules. The *Golgi apparatus* (named for Camillo Golgi, who first observed it) is a stack of flattened membranes to which many proteins made in the ER go for modification.

A cell can take on shapes radically different from spherical (for example, a sperm cell), and can change shape in response to changes in the environment. The shape of the cell is supported by the *cytoskeleton*, which, as its name implies, is the cell's structural framework. The cytoskeleton is composed of three major structural materials: microtubules, microfilaments, and intermediate filaments. *Microtubules* serve a number of functions. Among these are formation of the mitotic spindle—the apparatus that, during cell division, pushes one copy of each chromosome into each daughter cell. Microtubules are also the spine of eukaryotic cilia, which, like oars, can move the cell through its environment. Finally, microtubules can act as "railroad tracks" for molecular motors to carry cargo to distant parts of the cell. *Microfilaments*, thinner than microtubules, are made of the protein actin, which is also a major component of muscle. Microfilaments grab onto each other and slide to contract. This shapes the cell by folding the cellular membrane at the right places *Intermediate filaments*, which are thicker than microfilaments but thinner than microtubules, seemingly act simply as structural supports (like steel girders). Intermediate filaments are the most diverse structures of the cytoskeleton.

Almost all eukaryotic cells contain the organelles described above. Plant cells, however, contain several additional organelles. The *chloroplast* is the site of photosynthesis. Chloroplasts are, in many ways, similar to mitochondria since they both have energy-generating responsibilities. Chloroplasts contain the pigment chlorophyll, which acts as an antenna to catch light. The energy of the light is passed to extremely complex molecular machinery that generates differences in acidity across the membranes of the chloroplast. Plant cells also have a large, clear, membrane-enclosed space called the *vacuole*. The vacuole is a reservoir for wastes, nutrients, and pigments, and it also has a structural role. The vacuole occupies about 90 percent of the volume of some plant cells and is under high osmotic pressure. The pressure, pushing against a strong plant cell wall, stiffens the cell.

PROTEIN STRUCTURE

The cells and organelles described above, although quite tiny by everyday standards, are very large compared to the building materials of which they are composed. The building materials of cells and subcellular structures are ultimately composed of *atoms* stitched together into *molecules*. A chemical bond, or *covalent* bond, forms when each of two atoms contributes an electron to share between them. By sharing negatively charged electrons, the atoms more efficiently screen their positively charged atomic nuclei. A molecule is two or more atoms covalently bonded to each other.

Surprisingly, the types of atoms found in biological molecules are few. Almost all biomolecules are made of atoms of six elements: carbon (C), oxygen (O), nitrogen (N), hydrogen (H), phosphorous (P), and sulfur (S). Some other elements (such as chlorine, sodium, calcium, potassium, magnesium, and iron) are found as ions in biological systems. (Ions are electrically charged particles that float more or less freely in water.)

Atoms of C, H, O, N, P, and S can bond with each other. Carbon can bond with up to four different atoms at once, and biological phosphorus can also bond four different atoms (almost always four oxygens). Nitrogen can form three bonds (four in special cases), and oxygen and sulfur can form two. Hydrogen can form only one bond to another atom. Carbon is unique among the elements in that it can form stable bonds with other carbon atoms to form long chains. Since a carbon in the middle of a chain has used only two of its bonds—one to bond to the carbon on its right, and the other to bond to the carbon on its left—it still has two more bonds to make. It can use one to bond, say, a nitrogen atom and the other ·perhaps to bond to another chain of carbon atoms.

The number of molecules that can be built from carbon and the other biological elements is very large indeed. Biological systems, however, don't use a large number of completely different molecules. Rather, a limited number of molecules are made and the large, "macro" molecules of life—such as proteins, nucleic acids, and polysaccharides—are constructed by stringing together in different arrangements molecules from the limited set. This can be likened to making an enormous number of different words and sentences from the twenty-six letters of the alphabet.

The building blocks of proteins are called *amino acids*. The twenty different amino acids that compose virtually all proteins have a common structure. On the left side of the molecule is a nitrogen-containing group called an amine, and on the right, joined to the amine by a central carbon atom, is a carboxylic acid group (hence the name amino acid). Also attached to the central carbon, in addition to a hydrogen atom, is another group, called the side chain (Figure A–2). The side chain varies from one type of amino acid to another. It is the side chain that gives an amino acid its particular character.

Amino acids can be grouped into several categories. The first group contains hydrocarbon side chains (side chains with only carbon and hydrogen atoms). These side chains are oily, like gasoline, and prefer to avoid contact with water molecules. The next group is the electrically charged amino acids; there are three positively and two negatively charged members. Charged side chains prefer to be in contact with water. Another group is the polar amino acids. Polar molecules, although not fully charged, have partially charged atoms in them. This arises when one atom pulls more strongly on the electrons than its partner atom in a chemical bond, bringing the electrons closer to it. The atom with the lion's share of the electrons has a somewhat nega-

FIGURE A–2

(Top) Four amino acids. The amino acids differ only in their side chains. (Bottom) The four amino acids have been chemically joined. Proteins are long chains of many chemically joined amino acids.

tively charged character, while the atom with a deficiency of electrons has a partial positive charge. Interactions between positively and negatively charged side chains, and between the partially positively and partially negatively charged atoms of polar side chains, can be very important in the structure of proteins.

During the synthesis of proteins, two amino acids are chemically joined together by reacting the amino group of one amino acid with the carboxylic acid group of another to form a new group called a peptide bond (Figure A–2). The new molecule still has a free amino group at one end and a free carboxyl at the other end, so another amino acid can be joined by contributing its amino end to form another peptide bond. This process can be repeated indefinitely until a macromolecule, containing hundreds or thousands of amino acid "residues" (the part left after the chemical reaction joining two amino acids), has been formed. Such macromolecules are known as *polypeptides* or *proteins*.

A typical protein contains anywhere from about fifty to about three thousand amino acid residues. The amino acid sequence of a protein is called its *primary structure*. The completed protein still has a free amino group at one end, referred to as the N-terminal end, and a free carboxyl at the other end, called the C-terminal end. The amino acid sequence of a protein is conventionally written starting from the N-terminal to the C-terminal end. The atoms of the protein joined in a line from the N to the C terminal are called the protein *backbone;* this includes all atoms except those of the side chains.

A freshly made protein does not float around like a floppy chain. In a remarkable process, virtually all biological proteins fold up into discrete and very precise structures (Figure A–3) that can be quite different for different proteins. This is done automatically through interactions such as a positively charged side chain attracting a negatively charged side chain, two hydrophobic side chains huddling together to squeeze out water, large side chains being excluded from small spaces, and so forth. At the end of the folding process, which typically takes anywhere from fractions of a second to a minute, two different proteins can be folded to structures as precise and different from each other as a three-eighths-inch wrench and a jigsaw. And, like the household tools, if their shapes are significantly warped, then they fail to do their jobs.

When proteins fold, they do not flop together like a string crushed

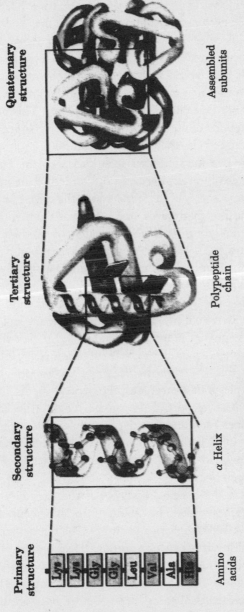

Primary structure

Secondary structure

Tertiary structure

Quaternary structure

Amino acids

α Helix

Polypeptide chain

Assembled subunits

Lys
Lys
Gly
Gly
Leu
Val
Ala
His

Copyright Irving Geis. Reproduced with permission

in your hand; there are regularities to the folding. Before a protein folds, its polar backbone atoms—the oxygen and nitrogen and hydrogen atoms in each peptide bond—form what are called *hydrogen bonds* to water molecules. A hydrogen bond occurs when a partially negatively charged peptide oxygen or nitrogen atom associates closely with the partially positively charged hydrogen atoms of water. When a protein folds, however, it must squeeze out all (or almost all) of the water so that the oily side chains can pack efficiently. This poses a problem: the polar peptide atoms must find oppositely charged partners in the folded protein, or else the protein will not fold.

There are two ways proteins solve this problem. First, segments of the protein can form an *α-helix*. In this structure the protein backbone spirals. The geometry of the spiral makes the oxygen atom of a peptide group point directly towards, and hydrogen bond with, the hydrogen of the peptide group found four amino acid residues back along the chain (Figure A–3). The next residue hydrogen bonds with the subsequent residue four back from it, and so on. Usually an α-helix has anywhere from five to twenty-five amino acid residues before the helical structure (but not necessarily the protein chain) ends. An α-helix permits a protein to fold into a compact shape while still forming hydrogen bonds to peptide atoms. A second structure that allows regular hydrogen bonding of peptide atoms is called a *ß-pleated sheet,* or simply a ß-sheet. In this structure the backbone of the protein goes up and down, like pleats in a sheet, and the peptide atoms stick out perpendicular to the direction of the protein chain. The chain then curls around, comes back, and the oxygen atoms in the peptide group of the returning strand hydrogen bond to the peptide group of the first strand. As with α-helices, ß-sheets allow polar backbone atoms to form hydrogen bonds.

α-helices and ß-sheets are known as the *secondary structure* of the protein. A typical protein has about 40 to 50 percent of its amino acid residues involved in helices and sheets. The remainder of the residues are involved in turns between portions of secondary structure, or else form irregular structures. Helices and sheets pack against each other to form, in most cases, a compact, globular protein. The exact way in which the elements of secondary structure pack is called the *tertiary structure* (Figure A–3) of the protein. The driving force for the packing of the helices and sheets comes from the oily nature of many protein side chains. Just as oil separates from water to form a distinct layer, so the oily, hydrophobic side chains huddle together to form a water-free zone in the interior of the protein. Recall, however, that some protein side chains are either polar or charged, and they want to stay in the water. The pattern of oily and polar side chains along the amino acid sequence, and the need for the protein chain to fold so that most of the hydrophobic groups are in the interior of the protein and most of the hydrophilic groups are on the exterior, provides the information that drives a specific protein to fold to a specific structure.

Another factor also contributes to the specificity of protein folding. In all folded proteins some polar side chains inevitably get buried. If the buried polar atoms do not find hydrogen-bonding partners, then the protein is destabilized. In most proteins about 90 percent of the buried polar side chain atoms are, in fact, hydrogen bonded to other side chains or to the protein backbone in a catch-as-catch-can manner. The folding of a typical protein—with its requirements to accommodate hydrophobic and hydrophilic groups and to form a network of hydrogen bonds—can be likened to a three-dimensional jigsaw puzzle.

Frequently, several separate polypeptides stick together in a very specific way to form a composite structure that functions as one entity. In these cases it is the custom to refer to the associated polypeptides as a single protein composed of several "subunits." For example, the oxygen-carrying protein hemoglobin is composed of four polypeptides, and the amalgamated protein has oxygen-binding properties that the component polypeptides lack. Thus the functional biological protein is the complex of the four polypeptides. The specific arrangement of separate polypeptides in a protein is called its *quaternary structure* (Figure A–3).

NUCLEIC ACID STRUCTURE

Like proteins, nucleic acids are polymers of a small number of building blocks, called *nucleotides*. A nucleotide itself has several parts. The first part is a carbohydrate, either ribose (in RNA) or deoxyribose (in DNA). To ribose is attached one of four *bases*, either adenine (A), cytidine (C), guanine (G), or uracil (U). If the carbohydrate is deoxyribose then U is replaced by a similar base called thymine (T); A, C, and G are also used with deoxyribose. Attached to a different part of the carbohydrate ring (to the 5'-OH or "five-prime hydroxyl" group) is a phosphate group. The sugar-phosphate portion of a nucleotide is analogous to the backbone portion of an amino acid, and the base is analogous to an amino acid side chain. It is only in its base that one nucleotide differs from another.

Two nucleotides can be joined chemically by reacting the phosphate of one nucleotide with the 3'-OH group of the carbohydrate portion of the second nucleotide (Figure A–4). This still leaves a free phosphate group on one end and a free 3'-OH group on the other end, which can be further reacted with other nucleotides. Repetition of this process can generate very long polynucleotides indeed. Cellular RNA ranges from about seventy to about fifty thousand nucleotides in length. One single molecule of DNA ranges from several thousand to about a billion nucleotides. The sequence of a polynucleotide is conventionally written starting from the 5' end to the 3' end.

Cellular RNAs are found as single polynucleotide chains. There are several biological classes of RNA. The first is called *messenger RNA* (mRNA); members of this class are produced as faithful transcripts of DNA genes; the genetic information carried by mRNA is then interpreted by the protein synthetic apparatus to produce a protein. The second type of RNA is called *ribosomal RNA* (rRNA). Polynucleotides in this class associate with a large number of different proteins to form the ribosome, the primary engine of protein synthesis. The last major category of RNA is called *transfer RNA* (tRNA). Members of this class are relatively small, seventy to ninety nucleotides in length, and serve as "adaptors" between the mRNA and the growing protein that is produced by the action of the ribosome.

Cellular DNA is found as a double-stranded molecule—two intertwined polynucleotides (the famous double helix) that are strongly held together by hydrogen bonding. To understand the reason for this we must look at the structure of the bases of the nucleotides (Figure A–4). The nucleotides can be divided into two categories—the purines (A and G), which carry the larger bases (composed of two fused rings), and the pyrimidines (C and T), which have only one ring. If A and T are correctly oriented, they can form two hydrogen bonds with each other, and G can form three hydrogen bonds with C. In cells, wherever there is a G in one strand of DNA there is a C in the second strand, and vice versa; and wherever there is an A in one strand there is a T in

FIGURE A–4

A PIECE OF DNA CONTAINING FOUR NUCLEOTIDES.

From Conn, E. E., Stumpf, P. K., Bruening, G., and Doi, R. H. (1987) *Outlines of Biochemistry*, 5th ed., John Wiley & Sons, New York, fig. 6.1. Reproduced with permission.

the second strand, and vice versa. Thus the two strands are called "complementary" to each other. To be correctly oriented for hydrogen bonding the two strands must be pointed in different directions, with one running 5' to 3' from left to right and the other going 5' to 3' from right to left. The DNA of eukaryotes consists of two complementary linear strands, but the DNA of many bacteria consists, surprisingly, of two complementary *circular* strands.

The amount of DNA in a cell varies roughly with the complexity of the organism. Bacteria have about several million nucleotides of DNA. The amount of eukaryotic DNA ranges from a low of several tens of millions of nucleotides in fungi to a high of several hundred billion in some flowering plants. Humans come in at around three billion nucleotides.

LIPIDS AND POLYSACCHARIDES

Two other major categories of biomolecules are lipids and polysaccharides. Polysaccharides are polymers of sugar molecules or their derivatives and play a variety of roles. They can be used as structural materials, such as the cellulose found in woody plants and trees, and as repositories of energy, such as the glycogen which is stored in the liver. Lipids, unlike proteins, nucleic acids, and polysaccharides, are not polymers made from discrete building blocks; rather, each lipid molecule must be synthesized from very basic starting materials. Lipids are not macromolecules, but they can associate to form large structures such as membranes.

TRANSCRIPTION

DNA, the repository of genetic information, is a polynucleotide. But the information it carries tells the cell how to make polypeptides—proteins. How does the information get translated from one polymer "language" to the other? Shortly after the discovery of the double helical structure of DNA physicist George Gamow proposed the very nonchemical idea that genetic information is stored in coded form, and that expressing the information involves decoding the polynucleotide and translating the message into the polypeptide language of proteins.[3] Although he was wrong about the specific nature of the code, Gamow's intuition was prophetic.

polymerase to become tightly overwound.[4] This would cause transcription to slow down or halt completely except that another protein, called *topoisomerase,* untangles the DNA. It does this by a complicated maneuver—cutting one strand of the tangled DNA, passing the uncut DNA strand through the cut strand, and then resealing the cut.

Transcription stops when the RNA polymerase runs into a special DNA sequence. In prokaryotes it is a palindromic[5] region containing about six or seven GC base pairs followed by a region of the same length rich in AT base pairs. Some, but not all, genes require an additional protein, called ρ, to make the polymerase fall off the DNA.

GENE REGULATION

A typical bacterial cell contains thousands of genes, and a typical mammalian cell contains tens of thousands. How does a cell know when to transcribe a gene, and how does it select a specific gene from the thousands available? The problem of "gene regulation" is a major focus of research. Many details have been uncovered, but much remains murky. One of the simplest examples of gene regulation is the regulation of the life cycle of bacteriophage λ. Bacteriophages—the prokaryotic analogs to viruses—are bits of DNA wrapped in a protein coat. In order to make copies of itself, a bacteriophage must find a suitable bacterial cell, attach itself to the cell, and inject its DNA into the host. The DNA from the phage is quite small, coding for only about fifty genes. This is not sufficient to make its own replication machinery so, cleverly, the phage hijacks the host's machinery. Thus the phage is a parasite, unable to provide completely for itself.

Sometimes when bacteriophage λ invades a cell, the cell makes so many copies of λ that it bursts. This is called the *lytic* cycle. At other times, however, λ inserts its own DNA into the bacterial DNA, making a single molecule from two. There the λ DNA can rest quietly, be replicated along with the rest of the bacterial DNA when the cell divides, and bide its time. This is called the *lysogenic* cycle. When the bacterium, perhaps many generations later, runs into trouble (by, say, encountering high doses of ultraviolet light), the λ DNA in the bacterial DNA switches to the lytic mode. Only now does the phage make thousands of copies of itself, bursting the cell and spilling out new bacteriophages.

What switches bacteriophage λ from the lysogenic to the lytic cycle? When bacteriophage DNA enters the cell, RNA polymerase binds to a bacteriophage λ transcription promoter. One of the first genes to be expressed is for an enzyme, called an "integrase," that chemically inserts the λ DNA into the bacterial DNA. The enzyme does this by cutting the circular λ DNA at a specific site that has a sequence similar to a site in the host DNA, which the integrase also cuts. This leaves both pieces of DNA with complementary, "sticky" ends that hydrogen bond to each other. The integration enzyme then joins the pieces of DNA.

Another λ gene codes for a protein called a "repressor." The repressor binds strongly to a sequence of λ DNA which RNA polymerase must bind to start the lytic cycle. When λ repressor is there, however, RNA polymerase cannot bind, so the lytic cycle is switched off. There are actually three binding sites for repressor—all in a row. Repressor binds the first site more strongly than the second site, and the second more strongly than the third. The third site overlaps the promoter for the gene that codes for the repressor itself. This arrangement allows the repressor to be synthesized continuously until the third site is filled, at which point synthesis stops. If the concentration of repressor falls to the point where it dissociates from the third site, then the repressor gene is again turned on.

By this mechanism λ repressor regulates its own production. In the presence of some chemicals, ultraviolet light, or other damaging agents, however, a gene for an enzyme that specifically destroys λ repressor is switched on. When the repressor is removed from the first site, the gene for a protein called Cro is activated. Cro protein binds strongly to the third λ repressor binding site, shutting it off forever, and launching the bacteriophage into the lytic cycle. All the genes necessary for making copies of the λ DNA and packaging them into protein coats are now transcribed.

The control of the life cycle of bacteriophage λ is one of the simplest examples of gene regulation. The regulation of other gene systems, especially in eukaryotes, can involve dozens of proteins. Nonetheless, it is thought that most genes are regulated by systems analogous to that of λ, with feedback controls and multiple factors conniving to decide whether a single gene should be turned on.

TRANSLATION

Once the messenger RNA has been produced, the task turns to translating the message into a protein. This process is best understood in prokaryotes.

The transcribed mRNA is bound by a particle called a ribosome. Ribosomes are huge complexes consisting of fifty-two separate proteins (of which several are present in multiple copies) and three pieces of RNA with lengths of 120, 1,542, and 2,904 nucleotides. The ribosome can be readily broken down into two large pieces, called the 30S subunit and the 50S subunit.[6] Incredibly, the ribosome is self-assembling. Experiments have shown that when ribosomes are separated into their components and then remixed, under the right conditions the components will spontaneously reform ribosomes.

The ribosome has a problem similar to that of RNA polymerase: the ribosome must find the point in the mRNA at which to begin translation. In prokaryotes the site is marked by a tract called the Shine-Dalgarno sequence, about ten nucleotides upstream from the initiation site. Initiation occurs at the first subsequent AUG sequence. (AUG codes for the amino acid methionine.) In eukaryotes, initiation usually begins simply at the first AUG sequence from the 5'-end of the mRNA.

Ribosomes cannot bind directly to mRNA by themselves; several other factors are required. In prokaryotes three proteins called *initiation factors*—labeled IF-1, IF-2, and IF-3—are necessary. To begin translation, IF-1 and IF-3 bind to the 30S ribosomal subunit. This complex then goes on to bind (1) to a previously-formed complex of a tRNA molecule carrying methionine and bound to IF-2, and (2) to the mRNA molecule at the initiation site. Next, the 50S ribosomal subunit binds to the growing complex, causing IF-1, IF-2, and IF-3 to fall off. In eukaryotes, translation initiation goes through similar steps, but the number of initiation factors can be as high as ten or more.

In the next step a second tRNA molecule, associated with a protein named elongation factor Tu (EF-Tu), comes in carrying the appropriate amino acid and binds to the ribosome. A peptide bond forms between the two amino acids held on the ribosome. The first tRNA molecule now has lost its amino acid, and the two covalently bonded amino acid residues are linked to the second tRNA. At this point the first tRNA dissociates from the ribosome, the second tRNA moves into the site on the ribosome previously occupied by the first tRNA, and the ribosome moves precisely three nucleotides down on the mRNA. This translocation process requires another protein called EF-G for some as-yet-unknown function.

These steps are repeated until the ribosome reaches a three-nucleotide sequence that corresponds to a stop codon. Another protein, called *release factor,* binds to the stop codon, preventing the ribosome from moving there. Additionally, the release factor changes the behavior of the ribosome. Instead of simply sitting on the mRNA waiting for the release factor to move, the ribosome cuts the completed polypeptide chain from the final tRNA molecule to which it is still attached, and the protein floats free into solution. The inactive ribosome then dissociates from the mRNA, floats away, and is free to begin another round of protein synthesis.

Other factors, too numerous to mention in this brief sketch, are also necessary for a functioning translation system. These include the enzymes that chemically place the correct amino acid onto the correct tRNA, various mechanisms to "proofread" the translation, and the role of chemical energy, in the form of the activated nucleotide GTP, at every stage of translation. Nonetheless, this outline may give the reader both an idea of the process by which genetic information is expressed and also an appreciation for the intricacies involved in that expression.

DNA REPLICATION

There comes a time in the life of every cell when it turns to thoughts of division. One major consideration in cell division is ensuring that the genetic information be copied and handed down uncorrupted; a great deal of effort is invested in that task.

In 1957 Arthur Kornberg demonstrated that a certain enzyme could polymerize the activated forms of deoxynucleotides into a new DNA molecule that was an exact copy of whatever "template" DNA Kornberg threw into the reaction mixture. He called the enzyme *DNA polymerase I* (Pol I). The scientific community was ecstatic about the find. Over the years, however, it has been shown that Pol I's primary role is not to synthesize DNA during cell division; rather, it is to repair DNA that has been damaged by exposure to ultraviolet light, chemical mutagens, or other environmental insults. Two other DNA polymerases, Pol II and Pol III, were later discovered. The role of Pol II remains murky: mutant cells lacking the enzyme exhibit no observable defects. Pol III has been identified as the major enzyme involved in DNA replication in prokaryotes.

DNA polymerase III is actually a complex of seven different subunits, ranging in length from about 300 to about 1,100 amino acid residues. Only one of the subunits does the actual chemical joining of nucleotides; the other subunits are involved in critical accessory functions. For instance, the polymerizing subunit tends to fall off the template DNA after joining only ten to fifteen nucleotides. If this happened in the cell the polymerase would have to hop back on hundreds of thousands of times before replication was complete, slowing replication enormously. However, the complete Pol III—with all seven subunits—does not fall off until the entire template DNA (which can be more than a million base pairs long) is copied.

In addition to a polymerizing activity Pol III possesses, ironically, a $3' \rightarrow 5'$ nuclease activity. This means that it can degrade polymerized DNA into free nucleotides, starting at a free 3' end and working back toward the 5' end. Now, why would a polymerase also degrade DNA? It turns out that the nuclease activity of Pol III is very important in ensuring the accuracy of the copying procedure. Suppose that the wrong nucleotide became incorporated into the growing DNA chain. Pol III's nuclease function allows it to step back and remove the incorrect, mispaired nucleotide. Correctly paired nucleotides are resistant to the nuclease activity. This activity is called "proofreading"; without it, thousands of times more errors would creep in when DNA was copied.

DNA replication begins at a certain DNA sequence, known appropriately as an "origin of replication," and proceeds in both directions at once along the parent DNA. The first task to be tackled during replication, as for transcription, is the separation of the two parent DNA strands. This is the job of the *DnaA* protein. After the strands are separated two other proteins, called *DnaB* and *DnaC,* bind to the single strands. Two more proteins are recruited to the growing "bubble" of open DNA: *single strand binding protein* (SSB), which keeps the two parent DNA strands separated while the DNA is copied; and *gyrase,* which unknots the tangles that occur as the complex plows through double stranded DNA.

At this point DNA polymerase can begin synthesis. But several problems arise. DNA polymerase cannot start synthesizing by joining two nucleotides the same way that RNA polymerase starts transcription; the DNA enzyme can only add nucleotides to the end of a preexisting polynucleotide. Thus the cell employs another enzyme to make a short stretch of RNA on the exposed DNA template. This enzyme can begin RNA synthesis from two nucleotides. Once the RNA chain has gotten to be about ten nucleotides long, the DNA polymerase can then use the RNA as a "primer," adding deoxynucleotides to its end.

The second problem occurs as the replication "fork" opens up. The synthesis of one strand of new DNA can proceed without difficulty; this is the strand that the polymerase makes as it reads the template in a 3'→5' direction, making a new strand in a 5'→3' orientation, as all polymerases do. But how to synthesize the second strand? If done directly, the polymerase would have to read the template in a 5'→3' direction and thus synthesize the new strand in a 3'→5' direction. Although there is no theoretical reason why this could not occur, no known polymerase synthesizes in a 3'→5' direction. Instead, after a stretch of DNA has been opened up, an RNA primer is made near the fork and DNA synthesis proceeds backward, away from the replication fork, in a 5'→3' direction. Further synthesis on this "lagging" strand must wait until the replication fork opens up another stretch of DNA; another RNA primer must then be made, and DNA synthesis proceeds backward toward the previously synthesized fragment. The RNA primers must then be removed, the gaps filled in with DNA, and the ends of the DNA pieces "stitched together." This requires several more enzymes.

The above description of prokaryotic DNA replication has been pieced together by the enormous efforts of a large number of laboratories. The replication of eukaryotic DNA appears to be much more complex, and therefore much less is known about it.

NOTES

Preface

1. Cameron, A. G. W. (1988) "Origin of the Solar System," *Annual Review of Astronomy and Astrophysics, 26,* 441–472.
2. Johnson, P. E. (1991) *Darwin on Trial,* Regnery Gateway, Washington, DC, chap. 5; Mayr, E. (1991) *One Long Argument,* Harvard University Press, Cambridge, MA, pp. 35–39.

Chapter 1

1. By *biochemistry* I mean to include all sciences that investigate life at the molecular level, even if the science is done in a department with another name, such as molecular biology, genetics, or embryology.
2. The historical sketch presented here draws mainly from Singer, C. (1959) *A History of Biology,* Abelard-Schuman, London. Additional sources include Taylor, G. R. (1963) *The Science of Life,* McGraw-Hill, New York; and Magner, L. N. (1979) *A History of the Life Sciences,* Marcel Dekker, New York.
3. Described in Weiner, J. (1994) *The Beak of the Finch,* Vintage Books, New York.
4. Darwin, C. (1872) *Origin of Species,* 6th ed. (1988), New York University Press, New York, p. 151.

5. A good summary of the biochemistry of vision can be found in Devlin, T. M. (1992) *Textbook of Biochemistry,* Wiley-Liss, New York, pp. 938–954.

6. For example, as the expected pattern left by speciation events that occurred in isolated populations.

7. Farley, J. (1979) *The Spontaneous Generation Controversy from Descartes to Oparin,* Johns Hopkins University Press, Baltimore, p. 73.

8. Mayr, E. (1991) *One Long Argument,* Harvard University Press, Cambridge, chap. 9.

Chapter 2

1. Mann, C. (1991) "Lynn Margulis: Science's Unruly Earth Mother," *Science,* 252, 378–381.

2. Eldredge, N. (1995) *Reinventing Darwin,* Wiley, New York, p. 95.

3. Eldredge, N., and Gould, S. J. (1973) "Punctuated Equilibria: An Alternative to Phyletic Gradualism" in *Models in Paleobiology,* ed. T. J. M. Schopf, Freeman, Cooper and Co., San Francisco, pp. 82–115.

4. Beardsley, T. "Weird Wonders: Was the Cambrian Explosion a Big Bang or a Whimper?" *Scientific American,* June 1992, pp. 30–31.

5. Ho, M. W., and Saunders, P. T. (1979) "Beyond Neo-Darwinism—An Epigenetic Approach to Evolution," *Journal of Theoretical Biology* 78, 589.

6. McDonald, J. F. (1983) "The Molecular Basis of Adaptation," *Annual Review of Ecology and Systematics* 14, 93.

7. Miklos, G. L. G (1993) "Emergence of Organizational Complexities During Metazoan Evolution: Perspectives from Molecular Biology, Paleontology and Neo-Darwinism," *Memoirs of the Association of Australasian Paleontologists,* 15, 28.

8. Orr, H. A., and Coyne, J. A. (1992) "The Genetics of Adaptation: A Reassessment," *American Naturalist,* 140, 726.

9. Endler, J. A., and McLellan, T. (1988) "The Process of Evolution: Toward a Newer Synthesis," *Annual Review of Ecology and Systematics,* 19, 397.

10. Yockey, H. (1992) *Information Theory and Molecular Biology,* Cambridge University Press, Cambridge, England, chap. 9.

11. Kaplan, M. (1967) "Welcome to Participants" in *Mathematical Challenges to the Neo-Darwinian Interpretation of Evolution,* ed. P. S. Moorhead and M. M. Kaplan, Wistar Institute Press, Philadelphia, p. vii.

12. Schützenberger, M. P. (1967) "Algorithms and the Neo-Darwinian Theory of Evolution" in *Mathematical Challenges to the Neo-Darwinian Interpretation of Evolution,* ed. P. S. Moorhead and M. M. Kaplan, Wistar Institute Press, Philadelphia, p. 75.

13. Kauffman, S. (1993) *The Origins of Order,* Oxford University Press, Oxford, England, p. xiii.

14. Smith, J. M. (1995) "Life at the Edge of Chaos?" *New York Review,* March 2, pp. 28–30.

15. Mivart, St. G. (1871) *On the Genesis of Species,* Macmillan and Co., London, p. 21.

16. Aneshansley, D. J., Eisner, T., Widom, J. M., and Widom, B. (1969) "Biochemistry at 100°C: Explosive Secretory Discharge of Bombardier Beetles," *Science, 165,* 61; Crowson, R. A. (1981) *The Biology of the Coleoptera,* Academic Press, New York, chap. 15.

17. Hitching, F. (1982) *The Neck of the Giraffe,* Pan, London, p. 68.

18. Dawkins, R. (1985) *The Blind Watchmaker,* W. W. Norton, London, pp. 86–87.

19. Eisner, T., Attygalle, A. B., Eisner, M., Aneshansley, D. J., and Meinwald, J. (1991) "Chemical Defense of a Primitive Australian Bombardier Beetle (*Carabidae*): *Mystropomus regularis,*" *Chemoecology, 2,* 29.

20. Eisner, T., Ball, G. E., Roach, B., Aneshansley, D. J., Eisner, M., Blankespoor, C. L., and Meinwald, J. (1989) "Chemical Defense of an Ozanine Bombardier Beetle from New Guinea," *Psyche, 96,* 153.

21. Hitching, pp. 66–67.

22. Dawkins, pp. 80–81.

23. Dawkins, pp. 85–86.

24. Darwin, C. (1872) *Origin of Species,* 6th ed. (1988), New York University Press, New York, p. 154.

25. Dawkins, R. (1995) *River Out of Eden,* Basic Books, New York, p. 83.

Chapter 3

1. A good general introduction to cilia can be found in Voet, D., and Voet, J. G. (1995) *Biochemistry,* 2nd ed., John Wiley and Sons, New York, pp. 1253–1259.

2. There are also other connectors in this system. For example, the contacts the dynein arm makes with the microtubule also serve as a connector. As mentioned previously, a system can be more complex than the simplest system imaginable, and the cilium is an example of such a system.

3. Cavalier-Smith, T. (1978) "The Evolutionary Origin and Phylogeny of Microtubules, Mitotic Spindles, and Eukaryote Flagella," *BioSystems, 10,* 93–114.

4. Szathmary, E. (1987) "Early Evolution of Microtubules and Undulipodia," *BioSystems, 20,* 115–131.

5. Bermudes, D., Margulis, L., and Tzertinis, G. (1986) "Prokaryotic Origin of Undulipodia," *Annals of the New York Academy of Science, 503,* 187–197.

6. Cavalier-Smith, T. (1992) "The Number of Symbiotic Origins of Organelles", *BioSystems, 28,* 91–106; Margulis, L. (1992) "Protoctists and Polyphyly:

Comment on 'The Number of Symbiotic . . .' by T. Cavalier-Smith," *BioSystems*, 28, 107–108.

7. A search of *Science Citation Index* shows that each paper receives an average of less than one citation per year.

8. A good general introduction to flagella can be found in Voet and Voet, pp. 1259–1260. Greater detail about the flagellar motor can be found in the following: Schuster, S. C., and Khan, S. (1994) "The Bacterial Flagellar Motor," *Annual Review of Biophysics and Biomolecular Structure*, 23, 509–539; Caplan, S. R., and Kara-Ivanov, M. (1993) "The Bacterial Flagellar Motor," *International Review of Cytology*, 147, 97–164.

9. Voet and Voet, p. 1260.

Chapter 4

1. A good general introduction to blood coagulation can be found in Voet, D., and Voet, J. G. (1995) *Biochemistry*, John Wiley & Sons, New York, pp. 1196–1207. For more detailed descriptions see any of the following: Furie, B., and Furie, B. C. (1988) "The Molecular Basis of Blood Coagulation," *Cell*, 53, 505–518; Davie, E. W., Fujikawa, K., and Kisiel, W. (1991) "The Coagulation Cascade: Initiation, Maintenance, and Regulation," *Biochemistry*, 30, 10363–10370; Halkier, T. (1991) *Mechanisms in Blood Coagulation, Fibrinolysis and the Complement System*, Cambridge University Press, Cambridge, England.

2. The suffix *-ogen* designates the inactive progenitor of an active molecule.

3. The word *factor* is often used during research when it is not certain what the nature of a substance under investigation is—whether protein, fat, carbohydrate, or something else. Even after its identity is pinned down, however, sometimes the old name continues to be used. In the blood-clotting pathway, all "factors" are proteins.

4. A gene is a portion of DNA that instructs the cell how to make a protein.

5. Doolittle, R. F. (1993) "The Evolution of Vertebrate Blood Coagulation: A Case of Yin and Yang," *Thrombosis and Haemostasis*, 70, 24–28.

6. The proteins involved in blood clotting are frequently referred to by Roman numerals, such as Factor V and Factor VIII. Doolittle uses that terminology in his article in *Thrombosis and Haemostasis*. For clarity and consistency I have used the common names of the proteins in the quotation.

7. TPA has a total of five domains. Two domains, however, are of the same type.

8. The odds are not decreased if the domains are hooked together at different times—with domains 1 and 2 coming together in one event, then later on domain 3 joining them, and so on. Think of the odds of picking four black balls from a barrel containing black balls and white balls. If you take out four at once, or take two at the first grab and one apiece on the next two grabs, the odds of ending up with four black balls are the same.

9. This calculation is exceedingly generous. It only assumes that the four types of domains would have to be in the correct linear order. In order to work, however, the combination would have to be located in an active area of the genome, the correct signals for splicing together the parts would have to be in place, the amino acid sequences of the four domains would have to be compatible with each other, and other considerations would affect the outcome. These further considerations only make the event much more improbable.

10. It is good to keep in mind that a "step" could well be thousands of generations. A mutation must start in a single animal and then spread through the population. In order to do that, the descendants of the mutant animal must displace the descendants of all other animals.

Chapter 5

1. Alberts, B., Bray, D., Lewis, J., Raff, M., Roberts, K., and Watson, J. D. (1994) *Molecular Biology of the Cell*, 3rd ed., Garland Publishing, New York, pp. 556–557.

2. Kornfeld, S., and Sly, W. S. (1995) "I-Cell Disease and Pseudo-Hurler Polydystrophy: Disorders of Lysosomal Enzyme Phosphorylation and Localization," in *The Metabolic and Molecular Bases of Inherited Disease*, 7th ed., ed. C. R. Scriver, A. L. Beaudet, W. S. Sly, and D. Valle, McGraw-Hill, New York, pp. 2495–2508.

3. Pryer, N. K., Wuestehube, L. J., and Schekman, R. (1992) "Vesicle-Mediated Protein Sorting," *Annual Review of Biochemistry*, 61, 471–516.

4. Roise, D., and Maduke, M. (1994) "Import of a Mitochondrial Presequence into P. Denitrificans," *FEBS Letters*, 337, 9–13; Cavalier-Smith, T. (1987) "The Simultaneous Symbiotic Origin of Mitochondria, Chloroplasts and Microbodies," *Annals of the New York Academy of Science*, 503, 55–71; Cavalier-Smith, T. (1992) "The Number of Symbiotic Origins of Organelles," *BioSystems*, 28, 91–106; Hartl, F., Ostermann, J., Guiard, B., and Neupert, W. (1987) "Successive Translocation into and out of the Mitochondrial Matrix: Targeting of Proteins to the Inner Membrane Space by a Bipartite Signal Peptide," *Cell*, 51, 1027–1037.

5. Alberts et al., pp. 551–651.

Chapter 6

1. Good introductions to the immune system can be found in Voet, D., and Voet, J. G. (1995) *Biochemistry*, 2nd ed., John Wiley & Sons, New York, pp. 1207–1234; and Alberts, B., Bray, D., Lewis, J., Raff, M., Roberts, K., and Watson, J. D. (1994) *Molecular Biology of the Cell*, 3rd ed., Garland Publishing, New York, chap. 23.

2. The cells are actually called B cells because they were first discovered in the *Bursa fabricius* of birds.

3. The cell goes to enormous trouble to splice together gene pieces—employing very complex machinery to align the ends properly and stitch together the pieces. Except in the case of antibody genes, however, the *reason* that "interrupted genes" exist at all is still a mystery.

4. Except for cells that make special classes of antibodies. I won't discuss that further complication.

5. Bartl, S., Baltimore, D., and Weissman, I. L. (1994) "Molecular Evolution of the Vertebrate Immune System." *Proceedings of the National Academy of Sciences, 91,* 10769–10770.

6. Farries, T. C., and Atkinson, J. P. (1991) "Evolution of the Complement System," *Immunology Today, 12,* 295–300.

7. Examples include: DuPasquier, L. (1992) "Origin and Evolution of the Vertebrate Immune System," *APMIS, 100,* 383–392; Stewart, J. (1994) *The Primordial VRM System and the Evolution of Vertebrate Immunity,* R. G. Landes Co., Austin; Sima, P., and Vetvicka, V. (1993) "Evolution of Immune Reactions," *Critical Reviews in Immunology, 13,* 83–114.

Chapter 7

1. RNA is made of the four nucleotides A, C, G, and U.

2. Several other simplifications will be used. The hydrogen atoms of the molecule will not be discussed or indicated in Figure 7–1. Hydrogen atoms for the most part just ride along with other atoms in the synthesis of AMP, so it really isn't necessary to pay attention to them to get the idea across. Additionally, double bonds and single bonds will not be distinguished, since we are only interested in connectivity.

3. Zubay, G., Parson, W. W., and Vance, D. E. (1995) *Principles of Biochemistry,* Wm. C. Brown Publishers, Dubuque, IA, pp. 215–216.

4. Although it was previously thought that this step did not require ATP, more recent work has shown that ATP is necessary for the reaction to go at physiological concentrations of bicarbonate. Voet, D., and Voet, J. G. 1995. *Biochemistry,* 2nd ed., John Wiley & Sons, New York, p. 800.

5. Hall, R. H. (1971) *The Modified Nucleosides in Nucleic Acids,* Columbia University Press, New York, pp. 26–29.

6. Orò, J. (1961) "Mechanism of Synthesis of Adenine from Hydrogen Cyanide Under Plausible Primitive Earth Conditions," *Nature, 191,* 1193–1194. It should be kept in mind that just the base adenine is made by reactions of ammonia and hydrogen cyanide. The nucleotide AMP is extremely difficult to produce under plausible early earth conditions, as noted in Joyce, G. F. (1989) "RNA Evolution and the Origins of Life," *Nature, 338,* 217–224.

7. Quoted in Joyce, G. F., and Orgel, L. E. 1993. "Prospects for Understanding the Origin of the RNA World," in *The RNA World*, ed. R. F. Gesteland and J. F. Atkins, Cold Spring Harbor Laboratory Press, Cold Spring Harbor, NY, p. 18.

8. Except by the degradation of ATP, which must be made from AMP in the first place.

9. Creighton, T. (1993) *Proteins: Structure and Molecular Properties*, W. H. Freeman and Co., New York, p. 131.

10. Alberts, B., Bray, D., Lewis, J., Raff, M., Roberts, K., and Watson, J. D. (1994) *Molecular Biology of the Cell*, 3rd ed., Garland Publishing, New York, p. 14.

11. Ferris, J. P., and Hagan, W. J. (1984) "HCN and Chemical Evolution: The Possible Role of Cyano Compounds in Prebiotic Synthesis," *Tetrahedron, 40,* 1093–1120. It should be kept in mind that the compounds described in this paper do not have the foundation attached.

12. Bloom, A. (1987) *The Closing of the American Mind*, Simon and Schuster, New York, p. 151.

13. Horowitz, N. H. (1945) "On the Evolution of Biochemical Syntheses," *Proceedings of the National Academy of Sciences, 31,* 153–157.

14. For consistency with other descriptions, I have switched the letters A and D in Horowitz's paper.

15. Kauffman, S. (1993) *The Origins of Order,* Oxford University Press, New York, p. 344.

16. Smith, J. M. (1995) "Life at the Edge of Chaos?" *New York Review,* March 2, pp. 28–30.

Chapter 8

1. The atmosphere of the early earth is now thought to have been quite different from the one Miller assumed, and much less likely to produce amino acids by atmospheric processes.

2. Dose, K. (1988) "The Origin of Life: More Questions than Answers," *Interdisciplinary Science Reviews, 13,* 348.

3. Shapiro, R. (1986) *Origins: A Skeptic's Guide to the Creation of Life on Earth,* Summit Books, New York, p. 192.

4. Cech won the Nobel prize for his work. The awarding citation alludes to the impact of Cech's work on origin-of-life studies. Cech himself, however, rarely mentions the origin of life in connection with his work.

5. Joyce, G. F., and Orgel, L. E. (1993) "Prospects for Understanding the Origin of the RNA World" in *The RNA World,* ed. R. F. Gesteland and J. F. Atkins, Cold Spring Harbor Laboratory Press, Cold Spring Harbor, NY, p. 19.

6. Joyce and Orgel, p. 13.

7. Although many statements within the scientific community's own journals and books are pessimistic, public statements to the news media tend to be of

the everything-is-under-control variety. University of Memphis rhetorician John Angus Campbell has observed that "huge edifices of ideas—such as positivism—never really die. Thinking people gradually abandon them and even ridicule them among themselves, but keep the persuasively useful parts to scare away the uninformed." Campbell, J. A. (1994) "The Comic Frame and the Rhetoric of Science: Epistemology and Ethics in Darwin's *Origin*," *Rhetoric Society Quarterly, 24,* 27–50. This certainly applies to the way the scientific community handles questions on the origin of life.

8. Schlesinger, G. and Miller, S. L. (1983) "Prebiotic Syntheses in Atmospheres Containing CH_4, CO, and CO_2," *Journal of Molecular Evolution, 19,* 376–382.

9. Niketic, V., Draganic, Z. D., Neskovic, S., Jovanovic, S., and Draganic, I. G. (1983) "Radiolysis of Aqueous Solutions of Hydrogen Cyanide (pH 6): Compounds of Interest in Chemical Evolution Studies," *Journal of Molecular Evolution, 19,* 184–191.

10. Kolb, V. M., Dworkin, J. P., and Miller, S. L. (1994) "Alternative Bases in the RNA World: The Prebiotic Synthesis of Urazole and Its Ribosides," *Journal of Molecular Evolution, 38,* 549–557.

11. Hill, A. R., Jr., Nord, L. D., Orgel, L. E., and Robins, R. K. (1989) "Cyclization of Nucleotide Analogues as an Obstacle to Polymerization," *Journal of Molecular Evolution, 28,* 170–171.

12. Nguyen, T., and Speed, T. P. (1992) "A Derivation of All Linear Invariants for a Nonbalanced Transversion Model," *Journal of Molecular Evolution, 35,* 60–76.

13. Adell, J. C., and Dopazo, J. (1994) "Monte Carlo Simulation in Phylogenies: An Application to Test the Constancy of Evolutionary Rates," *Journal of Molecular Evolution, 38,* 305–309.

14. Otaka, E., and Ooi, T. (1987) "Examination of Protein Sequence Homologies: IV. Twenty-Seven Bacterial Ferredoxins," *Journal of Molecular Evolution, 26,* 257–268.

15. Alexandraki, D., and Ruderman, J. V. (1983) "Evolution of α-and ß-Tubulin Genes as Inferred by the Nucleotide Sequences of Sea Urchin cDNA clones," *Journal of Molecular Evolution, 19,* 397–410.

16. Kumazaki, T., Hori, H., and Osawa, S. (1983) "Phylogeny of Protozoa Deduced from 5S rRNA Sequences," *Journal of Molecular Evolution, 19,* 411–419.

17. Wagner, A., Deryckere, F., McMorrow, T., and Gannon, F. (1994) "Tail-to-Tail Orientation of the Atlantic Salmon Alpha-and Beta-Globin Genes," *Journal of Molecular Evolution, 38,* 28–35.

18. Indeed, some proteins we have discussed in this book have sequences or shapes similar to other proteins. For example, antibodies are shaped similarly to a protein called superoxide dismutase, which helps protect the cell against damage by oxygen. And rhodopsin, which is used in vision, is similar to a protein found in bacteria, called bacteriorhodopsin, which is involved in the

production of energy. Nonetheless, the similarities tell us nothing about how vision or the immune system could develop step-by-step.

One would have hoped that finding proteins with similar sequences would lead to the proposal of models for how complex biochemical systems might have developed. Conversely, the fact that such sequence comparisons do not help us understand the origins of complex biochemical systems weighs heavily against a theory of gradual evolution.

19. I have counted in this category papers that are listed in the journal index under the titles "Molecular Evolution," "Protein Evolution," and some miscellaneous topics.

20. Kimura, M. (1983) *The Neutral Theory of Evolution*, Cambridge University Press, New York.

21. Kauffman, S. A. (1993) *The Origins of Order: Self-Organization and Selection in Evolution*, Oxford University Press, New York.

22. Selander, R. K., Clark, A. G., & Whittam, T. S. (1991) *Evolution at the Molecular Level*, Sinauer Associates, Sunderland, MA.

23. Cold Spring Harbor Symposia on Quantitative Biology (1987), vol. 52, *Evolution of Catalytic Function*, Cold Spring Harbor Laboratory Press, Cold Spring Harbor, NY.

24. Lehninger, A. L. (1970) *Biochemistry*, Worth Publishers, New York, p. 17.

25. Lehninger, A. L., Nelson, D. L., and Cox, M. M. (1993) *Principles of Biochemistry*, 2nd ed., Worth Publishers, New York, p. viii.

26. Lehninger et al. (1993), p. 244.

27. Conn, E. E., Stumpf, P. K., Bruening, G., and Doi, R. H. (1987) *Outlines of Biochemistry*, 5th ed., John Wiley & Sons, New York, p. 4.

28. Voet, D., and Voet, J. G. (1995) *Biochemistry*, 2nd ed., John Wiley & Sons, New York, p. 19.

29. To its credit, the Voet and Voet text contains a disclaimer at the beginning of the standard discussion of a Stanley Miller–like origin-of-life scenario, which states that there are "valid scientific objections to this scenario."

Chapter 9

1. Kauffman, S. A. (1991) "Antichaos and Adaptation," *Scientific American*, August, p. 82.

2. Kauffman, S. A. (1993) *The Origins of Order*, Oxford University Press, Oxford, England.

3. Detecting design in patterns of coin flips or other systems that do not physically interact is done in other ways. See Dembski, W. (1996) *The Design Inference: Eliminating Chance Through Small Probabilities*, Ph.D. dissertation, University of Illinois.

4. This is a judgment call. One can never prove that a particular function is the only one that might be intended—or even that it is intended. But our evidence can get pretty persuasive nonetheless.

5. It is hard to quantify design, but it is not impossible, and future research should proceed in this direction. An excellent start has been made by Bill Dembski in his dissertation (Dembski, 1996), which attempts to quantify the design inference in terms of what he calls the "probabilistic resources" of a system.

6. Dawson, K. M., Cook, A., Devine, J. M., Edwards, R. M., Hunter, M. G., Raper, R. H., and Roberts, G. (1994) "Plasminogen Mutants Activated by Thrombin," *Journal of Biological Chemistry, 269,* 15989–15992.

7. Reviewed in Gold, L., Polisky, B., Uhlenbeck, O. & Yarus, M. (1995) "Diversity of Oligonucleotide Functions," *Annual Review of Biochemistry 64,* 763–797.

8. Joyce, G. F. (1992) "Directed Molecular Evolution," *Scientific American,* December, p. 90.

9. Benkovic, S. J. (1992) "Catalytic Antibodies," *Annual Review of Biochemistry 61,* 29–54.

10. Dawkins, R. (1995) *River Out of Eden,* Basic Books, New York, pp. 17–18.

Chapter 10

1. Cited in Barrow, J. D., and Tipler, F. J. (1986) *The Anthropic Cosmological Principle,* Oxford University Press, New York, p. 36.

2. Barrow and Tipler, p. 36.

3. Paley, W. *Natural Theology,* American Tract Society, New York, pp. 9–10.

4. Dawkins, R. (1985) *The Blind Watchmaker,* W. W. Norton, London, p. 5.

5. Paley, pp. 110–111.

6. Paley, pp. 199–200.

7. Paley, pp. 171–172.

8. Paley, pp. 184–185.

9. Dawkins, p. 5.

10. Dawkins, p. 6.

11. Sober, E. (1993) *Philosophy of Biology,* Westview Press, Boulder, Co, p. 34.

12. Sober, pp. 34–35.

13. Sober, p. 35.

14. Sober, pp. 37–38.

15. Shapiro, R. (1986) *Origins: A Skeptic's Guide to the Creation of Life on Earth.* Summit Books, New York, pp. 179–180.

16. Miller, K. R. (1994) "Life's Grand Design," *Technology Review* February/March, pp. 29–30.

17. Dyson, J. F. (1966) "The Search for Extraterrestrial Technology" in *Perspectives in Modern Physics*, ed. R. E. Marshak, John Wiley and Sons, New York, pp. 643–644.
18. Crick, F. H. C., and Orgel, L. E. (1973) "Directed Panspermia," *Icarus, 19*, 344.
19. Futuyma, D. (1982) *Science on Trial*, Pantheon Books, New York, p. 207.
20. Miller, pp. 31–32.
21. Miller, p. 32.
22. Gould, S. J. (1980) *The Panda's Thumb*, W. W. Norton, New York.

Chapter 11

1. Shapiro, R. (1986) *Origins: A Skeptic's Guide to the Creation of Life on Earth*, Summit Books, New York, p. 130.
2. Dickerson's essay can be found in *Journal of Molecular Evolution, 34*, 277 (1992), and *Perspectives on Science & Christian Faith, 44*, 137–138 (1992).
3. The reformulated rule is essentially identical to what a peripatetic philosopher of science named Michael Ruse testified were the defining characteristics of science during the 1981 trial to determine the constitutionality of the Arkansas "Balanced Treatment for Creation-Science and Evolution-Science Act." Judge William Overton's opinion overturning the law relied heavily on Ruse's ideas. The opinion has been strongly criticized as inept by other philosophers of science. Many relevant trial documents are collected in Ruse, M., ed. (1988) *But Is It Science?* Prometheus Books, Buffalo, NY.

Judge Overton, echoing Ruse, wrote of science that: "(1) It is guided by natural law; (2) It has to be explanatory by reference to natural law; (3) It is testable against the empirical world; (4) Its conclusions are tentative; i.e., are not necessarily the final word; and (5) It is falsifiable (Testimony of Ruse and other science witnesses)." Overton's opinion was received with scorn by other philosophers of science. Philip Quinn wrote, "Ruse's views do not represent a settled consensus of opinion among philosophers of science. Worse still, some of them are clearly false and some are based on obviously fallacious arguments" (in Ruse, 1988, p. 384). Larry Laudan ticked off the problems: "Some scientific theories are well-tested; some are not. Some branches of science are presently showing high rates of growth; others are not. Some scientific theories have made a host of successful predictions of surprising phenomena; some have made few if any such predictions. Some scientific hypotheses are *ad hoc*; others are not. Some have achieved a 'consilience of inductions'; others have not" (in Ruse, 1988, p. 348). Laudan cited many exceptions to Overton's opinion: "This requirement [for explanation by natural law] is an altogether inappropriate standard for ascertaining whether a claim is scientific. For centuries scientists have recognized a difference between establishing the existence of a phenomenon and explaining that phenomenon in a lawlike way. . . . Galileo and Newton took themselves to have established the existence of gravitational phenomena, long before anyone was able to give a causal or explanatory account of gravitation. Darwin took himself to have established the existence of natural selection almost a half-century before geneticists were able to lay out the laws of heredity on which natural selection depended" (in Ruse, 1988, p. 354). Laudan saw no cause for rejoicing: "The victory in the Arkansas case was hollow, for it was achieved only at the expense of perpetuating and canonizing a false stereotype of what science is and how it works" (in Ruse, 1988, p. 355).

4. Of course, whether "evolution" and "religion" are compatible depends on your definitions of both. If one takes the position that evolution not only occurred solely by uninterrupted natural law, but that the process is "purposeless" and "unforeseen" in a metaphysical sense, then that does place "evolution" on a collision course with many religious denominations. Phillip Johnson has done an admirable job of pointing out the many ways in which the word *evolution* is used, and how shifting definitions can confuse public discussion of the issue. Johnson, P. E. (1991) *Darwin on Trial,* Regnery Gateway, Washington, DC.

5. Simon, H. (1990) "A Mechanism for Social Selection and Successful Altruism," *Science, 250,* 1665–1668.

6. The influence of various religious cultures on the development of science is described in Jaki, S. (1986) *Science and Creation,* Scottish Academic Press, Edinburgh.

7. The reaction of science to the Big Bang hypothesis, including Eddington's and other prominent physicists, is recounted in Jaki, S. (1980) *Cosmos and Creator,* Regnery Gateway, Chicago.

8. Jaki, S. (1986).

9. Dawkins, R. (1986) *The Blind Watchmaker,* W. W. Norton, London, p. 159.

10. Dawkins, R. (1989) *New York Times,* April 9, 1989, sec. 7, p. 34.

11. Maddox, J. (1994) "Defending Science Against Anti-Science," *Nature, 368,* 185.

12. Dennett, D. (1995) *Darwin's Dangerous Idea,* Simon & Schuster, New York, pp. 515–516.

13. Dawkins, R. (1986), p. 6.

Appendix

1. Prokaryotes can be subdivided into two categories: archaebacteria and eubacteria. The distinction does not matter for the present purpose of describing the internal architecture of cells.

2. Since cells are so small, visualizing them requires powerful microscopes. Most detailed "pictures" of cells are obtained by electron microscopy, in which electrons are used instead of light for illumination.

3. Gamow, G. (1954) "Possible Relation Between Deoxyribonucleic Acid and Protein Structure," *Nature, 173,* 318; Gamow, G., and Ycas, M. (1958) "The Cryptographic Approach to the Problem of Protein Synthesis," in *Symposium on Information Theory in Biology,* ed. H. P. Yockey, R. L. Platzman, and H. Quastler, Pergamon Press, New York, pp. 63–69.

4. The problem can be understood by the following example: Wind one shoe-string several times around another, and ask someone to hold the ends of the strings tightly in two hands. Now take a pencil, insert it between the strings near one hand, and push the pencil toward the other hand. The shoestrings in front of the moving pencil will become more tightly wound. The strings behind the pencil will be, in the jargon of biochemistry, "melted."

5. A palindrome is a word or sentence that reads the same way backward and forward. An example is "A man, a plan, a canal Panama." When applied to DNA, *palindrome* means a sequence of nucleotides that reads the same in the 5'→3' direction on both strands of the double helix.

6. The abbreviation S stands for Svedberg units, and is a measure of how fast a particle sediments in liquid.

ACKNOWLEDGMENTS

The development of this book has benefited greatly from conversations with many people. Many thanks to Tom Bethell and Phil Johnson for encouragement, and for showing this lab-bound scientist how to go about getting a book published. I'm grateful to my editor, Bruce Nichols, for saving the book from being a text chock-full of technical jargon, and for showing me how to arrange the pieces of the argument to make it more easily understood. I'd also like to thank Del Ratzsch and Paul Nelson for helping to firm up the argument, steering me past as many philosophical pitfalls as they could. Thanks to my Lehigh colleagues Linda Lowe-Krentz and Lynne Cassimeris for checking the science in the example chapters. I also appreciate the input of Bill Dembski, Steve Meyer, Walter ReMine, Peter van Inwagen, Dean Kenyon, Robin Collins, Al Plantinga, John Angus Campbell, and Jonathan Wells. The good points of the book are due to their help. Any deficiencies that remain are my own.

I'm glad to have the opportunity to publicly thank my wife, Celeste, for her unflagging support and encouragement, and for shouldering alone the happy but tiring task of running after our children while I spent evenings and weekends in the quiet of the office, pecking at the keyboard. I apologize to Grace, Ben, Clare, Leo, Rose, and Vincent for trips to the playground not taken and games of Frisbee not played. That will now change.

INDEX